the *Emotional* side of CONFLICT

A Practical Guide to the Science of Both

James Scott Harvey

East Prairie Lane Press

CONTENTS

INTRODUCTION

This book on emotions was written after years of study. It involved a long process of discovery starting as a basic search for information to support a work assignment. Curiosity gave way to interest in exploring and learning as much as possible. This was in the 1990's, widely considered an important decade on brain research due to the introduction of imaging technology. This medical technology provided a window to observe the function of the living human brain. The 1990's were called the Decade of the Brain due to all of the advances in research.[1] Through my own study, gradually the background story on emotions emerged into the main character. This observation led me into trying to apply what science has learned about emotions to the area of conflict resolution.

In an effort to describe emotions, this book assembles into a single narrative the published scientific work of many highly qualified researchers. Extensive work has gone into compiling current and accurate information on emotions by reading, distilling, and thinking about a wide variety of scientifically based brain research books and articles. Published authors with recognized expertise included Antonio Damasio, Richard Davidson, Lisa Feldman Barrett, Barbara Fredrickson, Daniel Kahneman, Stephen W. Porges, Eric Kandel, Daniel Schacter, Joseph LeDoux, Bruce Perry, Jill Bolte Taylor, James McGaugh, Torkel Klingberg, Bessel Van Der Kolk, David Eagleman, Daniel Goleman, and many more. Also reviewed were the publications from the Charles A. Dana Foundation, *Scientific American*, and much more.

In these published materials, each author was working to describe specific aspects of the human brain, including emotions. Each author had a very convincing case, presenting credible content based on science. All of this information and more was moving me in the right direction on understanding emotions. However, it seemed like a giant jigsaw puzzle with each author describing one piece.

The path used to tell the story weaves together into a single narrative a summary of the basic science of emotions, while following the evidence where it leads. The burden of proof is on the one making the claim. To address the burden of proof, the Endnotes section documents the primary sources used.

Emotions are not a single entity but a system doing a variety of functions across multiple layers of our body. We see the external surface (emotions expressed) supported by what is underneath. These underneath layers work, using smaller than microscopic circuits to create the emotions we experience as part of what makes our human brain the most complex structure in the known universe[2]. Decisions are influenced by the emotional system operating as unconscious, automatic, implicit, deeply hidden functions, which sway our rational choices[3]. This factual complexity can be summarized with the simple phase, "We have emotional brains that think."

The general purpose of the book is to report on the universal biological reality of human emotions in order to provide the reader with a solid foundation of knowledge on the science. The vision of the book is to put the science of human emotions into action in order to enhance the efforts needed to resolve disputes, solve problems, and improve relationships. In doing so, the hope is to make a positive difference in the reader's life by helping to improve the civic discourse within our community.

"For there is nothing either good or bad, but thinking makes it so."
—William Shakespeare[4]

I. THE CHALLENGE OF MANAGING OUR EMOTIONS

Time is short, and a decision must be made now. Our heart says one thing, but our head is leaning the other way. We like to believe there is some separation of the subjective emotional heart from the objective, thinking head. But in reality, our brain unconsciously and seamlessly blends what is in our thinking head with what is in our emotional heart. Our emotions are not only coloring the options available, influencing our thinking, but also are essential for making the decision. This is the story of the emotional side of conflict with the goal of putting science into action by improving awareness of the basic concepts involved.

Emotions are natural body functions needed to survive in the world. In order to better manage our emotions, we do need to know what they are. Over 350 years of neurobiological research has clearly established all mental functions come from the brain[5]. Thus, all conflict starts and ends in the brain, driven by emotions. It is not *if* a conflict will happen, but *when*. If one has knowledge of the universal biology of human emotions, when the conflict slips into dispute, we will have some additional useful tools to support the efforts required in the moment. This knowledge will also help us improve the important life skill of self-control.

Our emotions are constructed in the moment by core body functions, assisted by a lifetime of experience to engage in survival-oriented actions influencing thinking and behaviors. Our emotions mostly operate automatically and unconsciously. Built layer upon layer across bodily functions, including our brain, they seamlessly interconnect and function as an integrated whole. Biologically speaking, frequently we sense the effects from emotions before we think.

As a result, emotions influence a wide variety of processes, including attention, thinking, behaviors, memories, internal body functions, reading others' emotions, plus so much more. Our emotions include the natural biological functioning of unconscious hidden bias needed for making decisions. This emotional function biases us toward one option and against another, helping us to make the choices needed to navigate in the world.

What are basic concepts? Concepts are small units of knowledge stored in long-term memory.[6] As we encounter various contexts in living our lives, these concepts are used to classify the experience in some way, such as a negative threat, a positive opportunity, or a neutral situation. For teaching purposes, this book summarizes the basic concepts behind the emotional side of conflict.

II. EMOTIONS EXPLAINED

Emotions are natural universal biological functions needed to survive in the world while giving meaning to the experiences in life. Emotions begin as an unconscious automatic appraisal to start survival-oriented actions which in turn influence bodily functions, attention, thoughts, memories, feelings, and behaviors. Emotions work in the present to help shape the significance from our experiences, and influence decisions on what to do next.

Negative Emotions are more intense, narrowing and focusing our efforts in response to threats.[7]

Positive Emotions help to widen perspective and add to useful resources in response to opportunity.

Single-cell organisms have the capacity to address the challenges of living with no conscious consideration. They need to carry out all life functions independently, including responding to their environment from the threat of changes in light and temperature as well as the opportunity to find food.[8] Thus, built into every living cell is a desire for survival by responding to threats and opportunities, the core of negative and positive emotions. An opportunity for one may be a threat to another, as in the predator-prey relationship between a cat and mouse, or fox to rabbit. There are many words used to describe negative threats and positive opportunities.

Negative (–) Threat	Positive (+) Opportunity
Bad	Good
Pain	Pleasure
Dislike	Like
Stick	Carrot
Villain	Hero
Sadness	Euphoria
I hate it!	I love it!
Punish	Reward
Opposed	Enthusiasm
Failure	Success
Thumbs down	Thumbs up
Feeling vulnerable	Feeling safe
Adversarial	Collaborative
Discomfort	Comfort
Burden	Benefit
It rubbed me the wrong way.	I had a good feeling about it.

Emotional State is an unconscious biological chain reaction measured in seconds to minutes, using neural circuits plus the blood stream moving a variety of biochemical blends around the body to engage in short-term survival efforts.[9]

One cycle of an emotional state: 1. An incident happens; serves as a trigger for an unconscious automatic appraisal; 2. Our brain assesses the event as a threat or opportunity; 3. Automatically, biochemicals pour into bloodstream and nervous system; 4. This automatic release supports the actions required; 5. Negative or positive emotional response fully engaged; 6. In time the biochemicals are depleted by the efforts required; 7. "Fatigue" happens; and 8. Rest and digest to recover. Aftermath is the time period following the actions where the consequences become known and feedback is given.

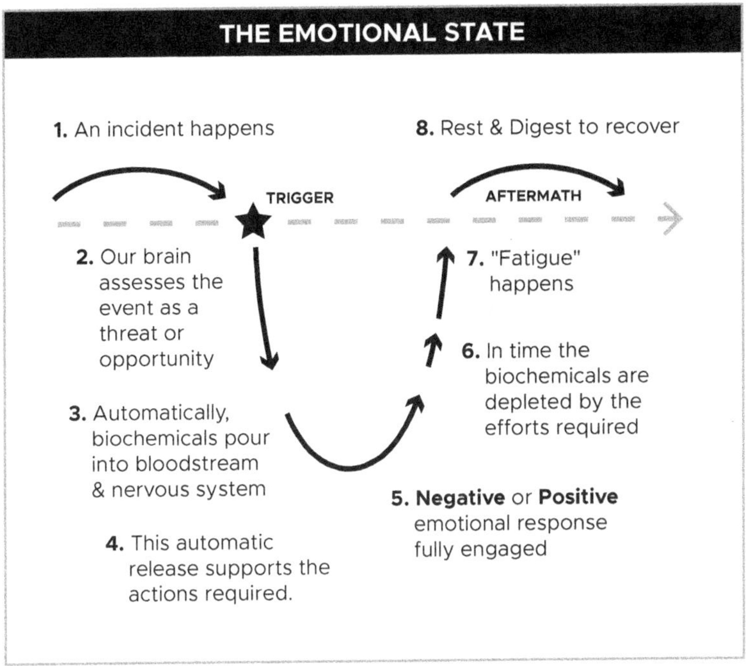

The variety of biochemical blends come with names like adrenaline, cortisol, oxytocin, serotonin, and dopamine.[10] They are part of the body's internal complex and nuanced messaging methods influencing our different responses to life events. Generally, there are hormones and neurotransmitters. The endocrine glands produce the hormones, which when released, flow through the bloodstream, influencing the target cells, tissues, and organs. Neurons linked together to form circuits, communicate using neurotransmitters. Neurons are specialized cells with the capacity to influence other cells. Where two neurons meet, there is a very small gap called the synapse. An electrical impulse traveling inside the neuron triggers the release of the chemical signal, sending it across the gap to the next nerve cell.

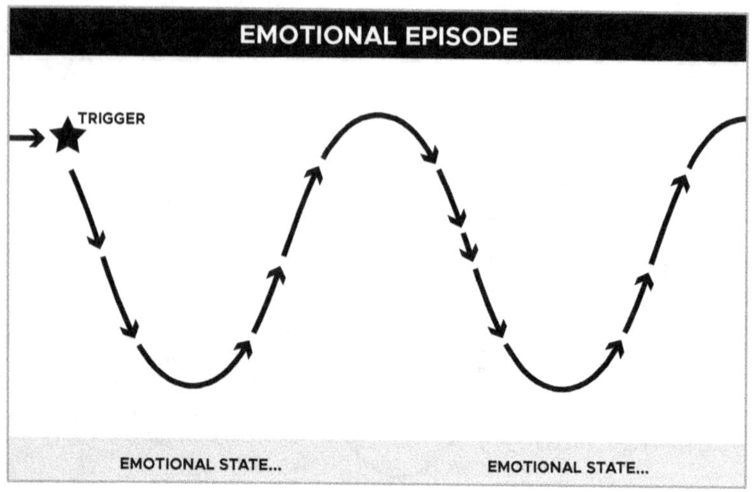

Emotional Episode involves the transition of one emotional state into another as the threatening stimulus or opportunity remains present. Basically, the emotional states keep flowing one into another as the situation unfolds.

Affect is our muscle movements for emotions observed by another person, such as facial expressions, tone of voice, posture, hand gestures, and other physical movements. Affect is thought of as our emotional "weather."[11]

Moods are persistent emotions coloring the perception of one's life remaining consistent for extended time periods measured in minutes, hours, or days. Because they are sustained and pervasive, moods are considered to be one's emotional "climate."[12]

Feeling Emotions involves sensing the biological changes inside our body during an emotional state. Feelings are the perceptions of the actual biological patterns supporting the emotional state. Feelings range from non-conscious sensing up to full conscious awareness with the ability to use words to identify the emotion.[13] A good use of this information is to start making the deliberate effort to notice these internal biological patterns in order to help categorize our experience as a negative and\or positive emotion.

Bias is the emotional system functioning effortlessly and automatically without any deliberate controls swaying our opinions, judgments, choices, attitudes, and decisions.[14]

Neuroception is a natural negative bias referring to the continuous, unconscious monitoring for dangers and threats.[15] Our emotional system is constantly doing an unconscious assessment. Is this a safe place? Is it okay for social engagement? We are cautious with our guard up, suspicious, and ready to withdraw until the all clear is established. If danger is sensed, natural biological defensive strategies can be triggered. Most living creatures work under the assumption of a constant threat of predators, setting up this natural negative bias. We fear before we trust. This leads to defensive actions in order to support our survival-oriented behaviors and thoughts. If sensed as safe, a damper on the vagus nerve inhibits the defensive strategies setting up the opportunity for social engagement. Natural biological defensive strategies include:

- *Freeze* (immobilization)—Shutdown, disengage, do not move because predators are better at noticing moving prey.

- *Fight* (mobilization)—Engage because our life is in imminent danger. We need to act with aggression.

- *Flight* (mobilization)—Run fast to escape using the safest path possible.

Traumatic Memory starts by directly experiencing or witnessing a dangerous life-threatening event.[16] The negative emotional event is formed and stored as a blend of implicit and explicit memory. Then, in the present moment, triggers cause parts of the past traumatic incident to be recalled, experienced as flashbacks. The reality with trauma is the scars in our mind long outlive the physical wounds healing. The emotional memory continues to play out in our body, actions, and thinking. There are many forms of life-threatening experiences seared into memory like Human-Caused Dangerous Events, Unintended Human Failure, and Natural Disasters.

Examples of Language Used to Label Emotional Concepts[17]

NEGATIVE EMOTIONS (narrow and focus our behaviors and thoughts to address threats.)

- FEAR—Threat, guard is up due to a sense of novelty or danger; fight, flight, freeze activated.

- ANGER—Violated, intense displeasure due to being damaged with desire for payback.

- SADNESS—Loss, grief, suffering because a positive part of our life is gone.

- DISGUST—Expel, desire to turn away due to aversion, distaste, revulsion.

POSITIVE EMOTIONS (widen perspective and add to our resources in response to opportunity.)

- GRATITUDE—Grateful. Acknowledge help from another; future duty to provide support.

- ELATION–Achieved a difficult challenge, accomplished the task.

- HOPE—A desire and expectation of a better future.

- AMUSEMENT—An activity providing pleasure and enjoyment.

- WONDER—In the presence of something overwhelming, extraordinary, or vast.

- INSPIRATION—Witness something influencing thoughts and emotions.

- LOVE—A safe relationship of strong affection for another person, place, or activity.

- JOY—A source of meaningful contentment; engaging, delightful, and happy.

III. THE EMOTIONAL SYSTEM

A system is an active group of interdependent functions performing vital tasks. The word *system* suggests connections, sequences, and interdependencies between parts of a unified whole. An active, ongoing, functioning system has inputs, processes, outputs, and feedback.[18]

The emotional system is an active dynamic group of interdependent body functions performing vital tasks to help us meet the demands of daily living needed for survival. There are various biological functions seamlessly integrated across our body used by the emotional system to carry out its own responsibilities.

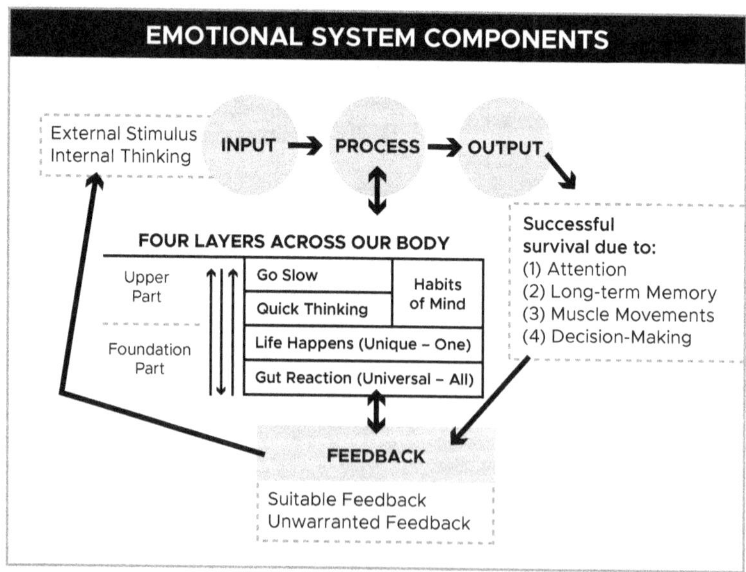

INPUTS—Live external stimuli received via the five outward-looking senses (sight, sound, touch, taste, smell), and internal thoughts in our head (something recalled or imagined). The internal function includes not only self-talk but also how these sensory stimuli are interpreted, considering the present, recalling the past, or imagining the future.

These inputs, referred to as triggers, start an unconscious automatic appraisal, using biological functions. A trigger is a cue or signal prompting action in the form of an emotional state. One strategy to improve self-control is to work on consciously noticing specific connections between the trigger and our emotional response.

PROCESSES—A process involves a series of vital tasks performed in a sequence leading to results. For teaching purposes our biological body can be divided into layers of emotional processes. These layers are seamlessly blended together and fully connected.[19]

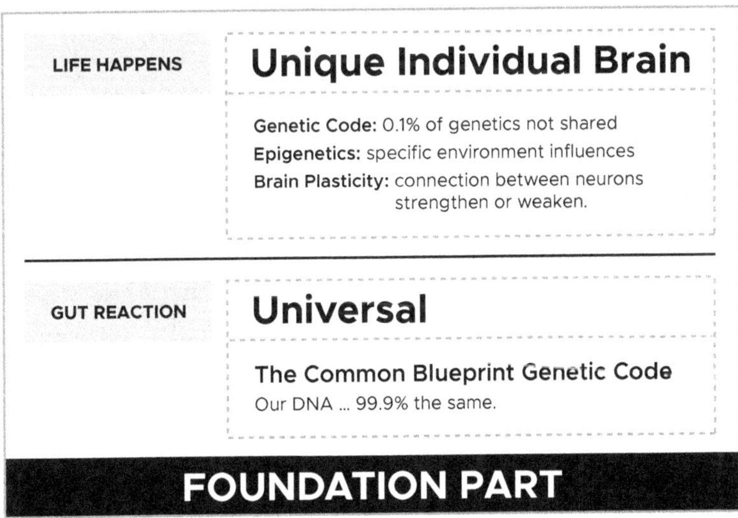

Foundation Part—Think about this as an unconscious biological structure like a foundation below ground level, holding up and supporting a building. The foundation part is two layers of non-verbal and unconscious biological supports needed for our body to function.

- **Gut Reaction** layer is the *universal* shared common biological blueprint hardwired into every person. It is based on the reality that all humans are 99.9 percent identical in our genetic makeup.[20]

  Key Gut Reaction functions are *homeostasis* and *interoception.*[21] Homeostasis involves the moment-to-moment biological processes needed to maintain a stable balance of resources within our body like oxygen, water, specific foods, and sleep under ever-changing conditions. Interoception is the brain's perception of how our body is currently functioning by reading these homeostatic changes. From these biological processes, we get a range of sensations from unpleasant to neutral to pleasant, a core part of emotions, setting up our gut reactions.

- **Life Happens** layer is *unique* to each individual. It starts with the *0.1 percent genetic makeup* specific to each individual. It is further shaped by the impact of the environment where the person lives. *Epigenetics* is the impact of the environment influencing gene expression.[22] *Plasticity* is our brain's capacity to change, based on lived experience by strengthening or weakening connections between neurons. This lived experience shapes the memory needed to adapt to our environment, helping us survive.

  Our attention short-term memory functions complete an appraisal of emotional meaning, which can provide a quick response to a dangerous threat or fine opportunity. Then selected parts of what we have heard, seen, tasted, smelled, touched, or thought in short-term memory goes through a consolidation process to form and store the content for later use. Significant emotional moments are given a higher priority. We better remember the way someone made us feel, especially if negative emotions are involved than what the person said.

  Unused portions of the memory are forgotten. It may feel insignificant in the moment, but the small changes flowing from these three elements add up over the course of a person's life with accumulative impacts. This is consistent with chaos theory on how small changes at nano level in an active living system like our individual body do add up over time to result in big differences.[23] These forces create our unique brain.

- **Universal Yet Unique** as a feature of the foundation part:

  → The gut reaction layer involves the universal biological basics of life, meaning all people share the same physical needs including food, water, air to breathe, and shelter for a safe place to sleep. None of us asked to be born. We will all die. Nobody is perfect. We are all one human family living on Planet Earth.

  → With the life happens layer, there are unique individual elements giving each of us a one-of-a-kind distinctive brain. Because each brain is unique, we each experience negative threats and positive opportunities differently; we start at different places with understanding, go at our own individual pace, follow our own pathway and tell our own stories.

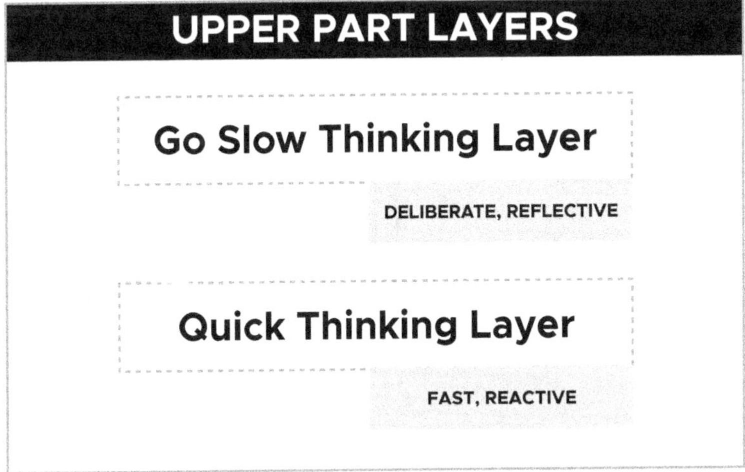

Upper Part—Think about these as two layers of conscious mental actions like one is upstairs in the living area of the building as seen from the street. Here, our thinking provides a sense of mental actions, which can be verbally expressed. The upper part includes routine ways of thinking called habits of mind as influenced by our Quick Thinking and Go Slow Layers.[24]

- **Quick Thinking** is automatic, fast, instinctive, reacting first without considering the actions done or the words spoken. It is nonconscious because the words can be spoken without much thought or awareness. Quick Thinking can seem like we are cruising with the ease of auto pilot. This includes the feeling of emotions, while not being consciously aware of them.

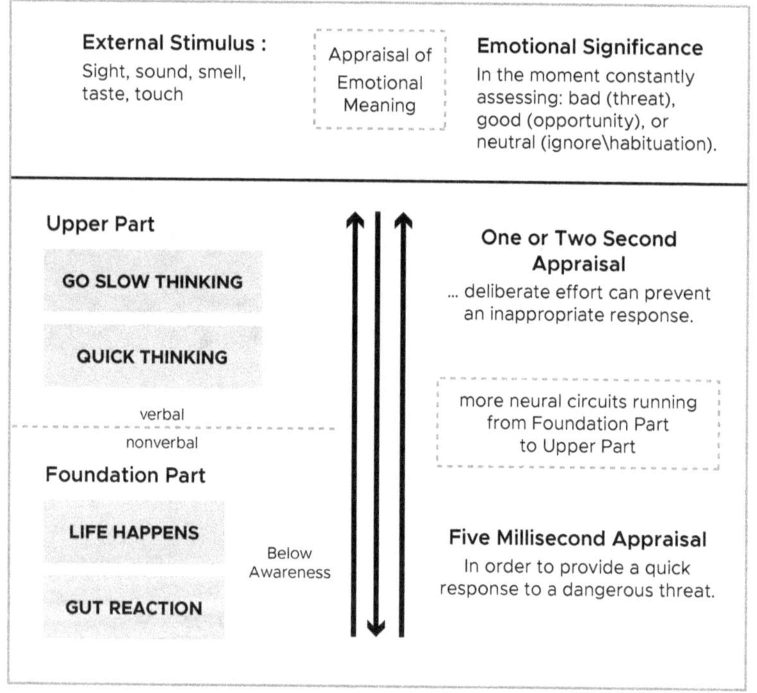

- **Go Slow Thinking** is conscious deliberate reasoning with effort. This layer has the capacity to make conscious choices, providing the basis for self-monitoring. The Go Slow Thinking effort is needed to counter what can feel like the instinctive suggestion from Quick Thinking, if it is not too late. The Go Slow Thinking is the place where we consciously notice what is going on around us, making it the home of our adult in charge. It is the sunlight zone where considered, measured, deliberate, slower, effort needed thinking with strain is done. Go Slow Thinking is responsible for sorting out truth from fiction.

Emotions start at the nano level, using those specialized cells called neurons.[25] The location where one neuron connects to another is called the synapse. At this connection point, the neurons do not actually touch. The space between the neurons is measured in nanometers (one billionth of a meter). These two neurons communicate, using electricity and chemicals. One neuron connected to another neuron at the synapse form into tiny microscopic threads to create a neural circuit. These microscopic threads in turn build into intricate webs of neural circuits seamlessly integrated to support the central and peripheral nervous systems.

The unconscious foundation part quickly processes signals before the thinking sections of the upper part get involved. The thalamus sitting at the upper part of the spinal cord is a key switching station. It quickly relays crude sensory impulses (vision, hearing, touch, taste; not smell) to the amygdala for a quick and dirty read of the situation. Smell goes directly from inside the nose to where it is processed in the brain.[26] One of the amygdala's duties is to store emotional memory. The amygdala uses this memory to quickly assess emotional significance and is good at noticing things coming through the senses as novel. If the quick and dirty read of a situation warrants it, the foundation part can place our body on alert or go ahead triggering an emotional state. Yes, this is a threat (negative emotion) or an opportunity (positive emotion). These foundation part circuits are very fast, taking about five milliseconds.[27]

At the same time, the thalamus sends these sensory impulses for further review by the outer layer of the brain called the cerebral cortex, the home of the of the upper part. There are more neural circuits running from the amygdala to the cortex than the other way around. This means the amygdala's emotional memory has more capacity to influence our conscious thoughts than our aware thinking gets to impact the foundation part. The upper part processing time can take an additional second or two.[28] While slower, the upper part can override the foundation part if it is not too late.

Sometimes we become consciously aware of this process. For example, most of us have experienced being startled while focused upon something. We are sitting concentrating on a task when out of the blue, someone comes along and says our name. Initially, in reaction to the unexpected sound, we are surprised, maybe jumping in alarm but then, a second or so later, we recognize our friend. We might say something like, "You just scared the daylights out of me." Our friend says sorry, and the discussion goes from there. That gap of five milliseconds with our initial startled reaction followed a second or so later of recognizing our friend is the processing time for the foundation part and upper part. It is all part of what makes our human brain the most complex structure in the known universe.

Habits of Mind are unconscious patterns of thought providing a way of processing information leading to routine ways of thinking shaped by experiences and reality. These are the mindsets we use to successfully deal with various pressures in our life. Habits of Mind are a blend of Quick Thinking and Go Slow Thinking, shaped over time. In the moment, they are influenced by emotions. Habits of Mind are unique to the individual.

Habits of Mind are our starting frame of reference helping to unconsciously organize patterns of thought within the context of the moment. They are a product of how the brain has been conditioned to think over our lifetime, based on experiences as well as responding to the natural urges and drives flowing from the foundation parts (gut reaction and life happens).

Our Habits of Mind seamlessly blend our nonconscious, Quick fast automatic Thinking and the conscious Go Slow deliberate Thinking as we routinely use them.[29] Thus the term habits of mind needs to be plural. As adults we have a wide variety of habits of mind to quickly deal with various situations. There can be a singular habit of mind like our slow, deliberate approach of using self-restraint to carefully think certain new and novel challenges through first. Then we use our nonconscious fast automatic ways of thinking in response to the same situations as they become routine. After we get good at what we practiced, these routines help us to conserve energy.

OUTCOMES—The emotional system results are to help us survive while giving meaning to the experiences in life influencing attention, long-term memory, muscle movements, and decision-making.

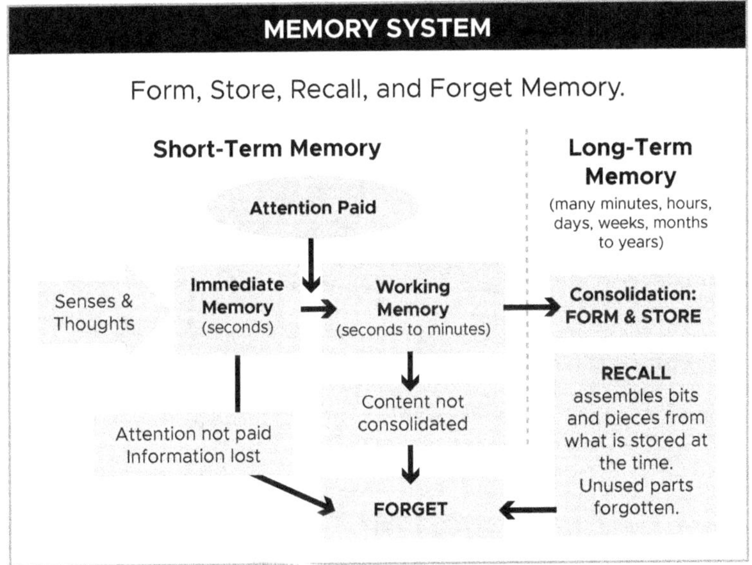

- **Attention** *(Short-Term Memory)*—Appraisal in the moment using filtering actions influenced by our emotions to narrow down to the current point of focus.[30]

  → The attention filters include: *neuroception* (natural negative bias of the continuous, unconscious monitoring for threats); *sensitization* (having a stronger reflexive response to certain stimulus based on past experience); *novelty detection* (something new and uncertain encountered); *reward pathway* (opportunity encountered); *habituation* (learning to ignore neutral non-threatening stimulus based on past experience); *bodily needs* (certain urges to address physical requirements like hunger or thirst); *emotional memory* (the important negative or positive experiences formed and stored in the past are now recalled in the moment); and *self-talk scripts* (narratives we tell ourselves).[31]

→ Distraction challenges include:

- *Attentional Blink* when one item captures our attention for the moment while related items of interest are ignored because the brain is too busy on processing the initial piece; and

- *Inattention Blindness* where we simply do not recognize the significance of the content directly in our gaze.[32]

- **Long-term Memory**—Emotional memories formed and stored in the past are recalled influencing actions right now in the present.

These functions can be summarized as:

Form	encoding, consolidating short-term memory with emotional significance into bits and pieces for long-term storage
Store	engram, a pattern of connections linking the bits and pieces, set up to last for a while
Recall	cues trigger retrieval of some bits and pieces assembled in the moment for use now
Forget	over time losing bits and pieces of the memory along the way

The physical pattern of the stored memory in the brain is called an engram, the enduring change to the physical, biological tissues needed to store the experience for the long-term.[34] Our long-term memory systems are organized to pull out of current events the useful information needed to cope with future situations.[35]

The useful information may be the patterns of factual, emotional, sensory (sight, sound, smell, taste, touch) and\ or the content in our current internal thoughts. The stronger the emotional component is, the stronger the memory will be. Long-term memory works to hold onto the main points of the experience unconsciously, asking what is key to retain from what just happened. The memory of the experience is organized from general content down to specific key elements in separate places (like visual features in one place, audio parts in another) linked together. This means the content of the experience gets stored in bits and pieces, as in a pattern of neural circuit connections now supporting the long-term memory, which can last for hours, days, weeks, months, up to years.

Typical long-term memory challenges include[36]

Lost	We forget bits and pieces over time.
Faulty	The bits and pieces stored are inaccurate.
Durable	The neural circuits supporting the bits and pieces are strongly maintained, as with traumatic events.

- **Muscle Movements**—Emotions influence the specific actions of our current body movements. This includes reading the muscle movements of other people's emotional states. The group's culture provides implicit agreements on what certain gestures and physical activity mean.[37] Some refer to this as affect, as in our muscle movements, as observed by another person like facial expressions, tone-of-voice, posture, hand gestures, and other physical body movements.[38] Looking at each other's facial expressions is so important that inside our brain, there is a specialized area with this assignment.[39] When this function does not work correctly for a few people, it is generally referred to as face blindness, a medical condition called prosopagnosia.[40]

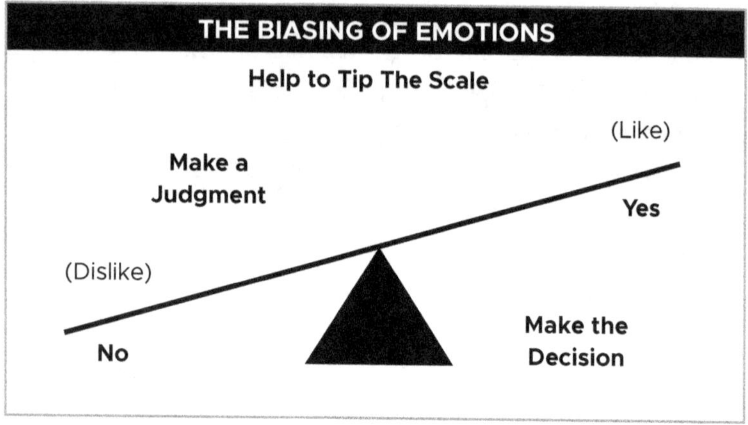

- **Decision-Making**—Emotions give meaning in the moment biasing the options available before choosing.

Bias is an important feature of decision-making. There are biological processes automatically setting up emotional content to sway the choices made which in turn impacts attention, long-term memory, and muscle movements. Emotional content is needed in order to place subjective value on the options being considered skewing the many dimensions down to a few possibilities and then just one. The prefrontal cortex integrates these signals from bodily responses and emotional memory in order to make decisions, form opinions, and pass judgments. Now the choice can be made.[41]

Thus, our rational thinking used for objective decision-making needs the biasing features of emotions. However, bias does hide things from our awareness by suppressing or ignoring certain stimuli. The importance of these biasing features can lead to the conclusion we have emotional brains that think.

FEEDBACK is there to help support, regulate, adjust, and balance the system. Feedback provides very important information to the emotional system on what is working and what is not, then helping to make the adjustments as needed.

An important feature to the lessons learned from our lived experience is feedback. Lived experience includes all kinds of mistakes, errors, accidents, and failures. At first approach, we do not know much of what is expected and mistakes are made. Over time, we move from these initial failures to some modest success, to improvement, onto mostly the expected performance, and the occasional perfection. The filtering actions of the attention functions within short-term memory do influence our current point of focus including what feedback to use or ignore. The key to the improvement is using the appropriate suitable feedback while ignoring the unwarranted comments.

→ *Suitable Feedback* is constructive criticism needed to ground us in the realities of life and face up to the facts. It is understandable, useful, and necessary, providing opportunity for course corrections as needed. Immediate corrective comments as part of appropriate feedback are an important feature of learning and building relationships. [42]

→ *Unwarranted Feedback* is excessive, unnecessary, disproportionate, and redundant under the circumstances.

Gotcha moments are deliberate efforts made as part of unwarranted feedback. We are focused on looking for something, which can appear wrong, disturbing, or embarrassing to the audience. When found, we proclaim got you—gotcha! Once caught, the action can be quickly shared with this individual and others. The shared thought sounds like, "I knew someone like you would do something like that!"

IV. THE IMPLICIT BIAS PHENOMENON

A key feature of the emotional side of conflict is the phenomenon of implicit bias, the end product of a long process. Bias is the natural, universal result of the four outputs of the emotional system. The implicit element of bias suggests biological processes working below our awareness automatically, involuntarily, without any deliberate control flowing at nano level. Within the foundation part, gut reaction and life happens levels complete an unconscious assessment with the filtering actions swaying attention on threat versus opportunity while associated memory gets recalled.

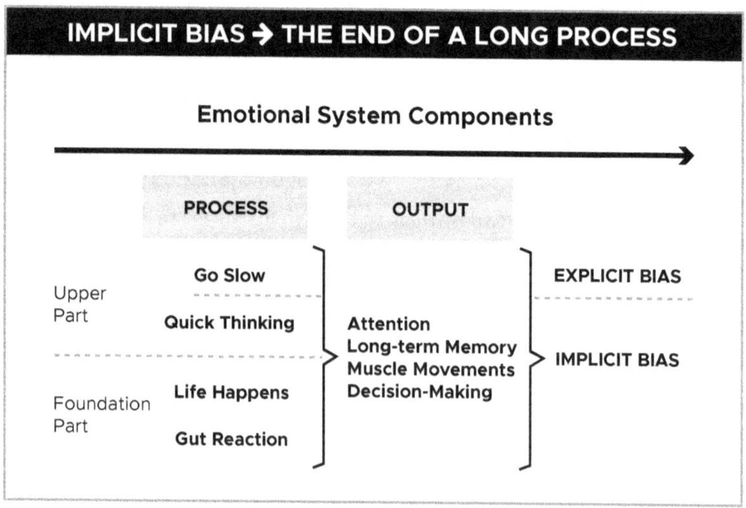

Implicit and Explicit Explained

Implicit means implied, informal, and unexpressed influences.[43] As a biological human feature implicit functions operate unconsciously, automatically, and reflexively. There are a variety of implicit influences that never reach our awareness.

- *Implicit Teaching* Methods are implied, wordless lessons demonstrated by leaders, role models, and seasoned group members in unexpressed ways. The students are expected to imitate the unsaid implied details. We are not conscious of these implicit interactions; however, over time they form, mold, and shape our thinking. This in turn influences our actions and behavior. Many cultural features are taught using implicit means. Then we get good at what we practice with the full support by those around us. It is also referred to as implicit conditioning.

- *Implicit Memory,* also referred to as non-declarative memory, is managed by a brain structure called the amygdala.[44] There is an amygdala on the left side of the brain and another on the right. The plural is amygdalae. Implicit memory operates automatically and reflexively, providing unconscious recall that never reaches our awareness. Emotional memory is unconscious implicit non-declarative content stored via the amygdala. While unaware of implicit *emotional memory* recall, it does influence our actions. It includes procedural memory, habituation, sensitization, priming cues, puzzle-solving skills, classical conditioning, and building associations between memories. For example, long after the physical wound heals, this type of emotional memory continues to unconsciously influence our choices.

- *Implicit Bias* is the effortless automatic emotional part working deeply hidden below conscious awareness without any deliberate controls. All of us have implicit bias in the moment as a natural inclination not obvious or apparent to us, unconsciously swaying our judgments, attitudes, and decisions.

Explicit means something is fully revealed and perfectly clear in its meaning. As a biological human feature, explicit functions usually suggest the focus of interest reaches our awareness setting up the opportunity for a verbal description.

- *Explicit Teaching* involves the deliberate work of presenting specific content in a classroom or related learning platform. It is also referred to as explicit conditioning.

- *Explicit Memory* is also referred to as declarative memory. Explicit memories are stored via a key brain structure called the hippocampus.[45] There is a hippocampus on the left side of the brain and on the right. The plural is hippocampi. Explicit memory is a depository of material about our personal history of places, people, and daily events of everyday life. Memory of emotional experiences are conscious, explicit declarative memories stored via the hippocampus used to verbally recall the events.

- *Explicit Bias* means we are able to verbally say the words and take the actions supporting our emotional point of view.

Our biases enter the upper part by Quick Thinking, our nonconscious ability for fast, automatic, reactive thoughts, with ease. It gives us the capacity to verbally talk without thinking much about what we are saying, as when expressing an opinion. That can be very handy in demanding moments with the automatic reaction without consideration. This sense of ease with Quick Thinking is important to help to save energy because our brain is about 2 percent of our body weight but uses 20 percent blood flow.[46] The blood flow serves as a proxy measure for the brain's energy use. The effortless features of implicit bias unconsciously influence our opinions and choices in the moment. Meanwhile, like the top of an iceberg, explicit bias floats to the highest part where we may be consciously aware of verbally saying the words and taking the actions supporting our point of view.

People important to us positively reinforce some preferences while certain opinions get roundly criticized by others. Afterwards,

Go Slow Thinking needs to explain what we did, rationalizing the actions taken and words spoken.

Thus, implicit bias is produced by emotional system processes and outputs:

(1) It is how emotional memory is formed, stored, recalled, and forgotten;

(2) The attention filters sway our focus;

(3) Our behaviors and actions are reflected in muscle movements, supporting body functions like facial expressions as well as tone of voice; plus

(4) Judgments, including opinions, are expressed and decisions made. They reflect our emotional preferences developed over time, such as likes and dislikes.

The whole package supporting the implicit bias becomes part of our habits of mind within the brain's routine interpretation of incoming information. So, disentangling the implicit bias as another emotional surge washes over us is very challenging. While we are unaware, our biases are publicly displayed for others to observe as simple preferences as well as stereotypes and overt prejudice. It takes paying attention to the feedback we receive from others to build the awareness of our own implicit bias. This is hard to do, but if we are aware of the implicit bias, we can work to reduce its impact by making the deliberate efforts to change.

Awareness of implicit bias is an important step to reducing its adverse impacts of what has been learned by a pattern of frequently repeated messages.[47] Go Slow Thinking needs to engage in the override option. It takes time and effort, but changes can be made. After all, opinions do change if new perspective can be gained.

A few other examples of bias include:

- *Affinity Bias*—As part of the us-versus-them dynamic, we find attractive features in other people who are similar to ourselves.[48]

We are not going to be automatically inclusive due to the neuroception bias. The unconscious bias does a threat assessment, quickly categorizing people by skin color, gender, height, weight, age, and more. Is this person safe? The natural filtering action creates gaps, as we may perceive other people like ourselves to be safe while those who are different a threat. In the moment, we sense how another person who agrees with us seems to be especially good, which in turn unconsciously influences our choices. The skew becomes self-maintaining as fewer of "thems" seek membership to be inside our group.[49]

- *Confirmation Bias*—is the inclination to attend to information consistent with our own preconceived notions. The unconscious bias narrows our focus to information supporting our own ideas and ignoring or discounting evidence to the contrary. This is done by a top-down filtering action that interprets the incoming stimuli. We are highly critical of facts not supporting our position. We stubbornly maintain how right we are and how wrong the other person is, even in the face of overwhelming contrary evidence.[50]

- *Tunnel Vision Bias*—In tunnel vision, one is narrow-minded due to tightly focused eyesight, excluding much of the peripheral content.[51] That old definition for tunnel vision covers the filtering action of seeing. Tunnel biasing effect takes the same idea but also includes the other senses of hearing, taste, smell, and touch. This filtering of the bottom-up sensory signals will be interpreted consistent with the current emotional state.

- *Willful Blindness*—We are in denial of an unpleasant reality by using specific efforts to deliberately avoid certain knowledge. The result is one intentionally stays selectively ignorant of facts supporting another point of view.[52]

V. REWARD PATHWAY

The reward pathway associates useful behaviors with pleasure, such as seeking food or finding a mate. After the reward has been consumed, time passes, and we seek to do it again because the positive effects do not last. Basic motivation has us either moving away to avoid a threat or approaching an opportunity. Reward is about the moving toward the opportunity. Thus, it is there to motivate us to look for opportunities and to seek out the more enjoyable positive things in life.

The biological function for this is the reward pathway.[53] This is where the pleasure circuits in our brain note the things we like or want.[54] We then get positive reinforcement for engaging in the rewarding behaviors. We also remember it in order to repeat the actions again. We get natural concrete, pleasurable feelings from eating good-tasting food, being nurtured, nurturing others, a chance for sexual activity, simply drinking water, and more.[55] This reward pathway can operate with various levels of abstraction, providing us with the pleasure of artistic pursuits, making money, political action, and other more intangible efforts. It may also be used to help reduce painful feelings after life's stresses and strains. These behaviors are needed for our survival. There are a variety of other structures in play during a pleasurable emotional state, such as memory so we may do the behavior again.[56]

Reward pathway brain structures connected by neural circuits include: the ventral tegmental area (VTA), the nucleus accumbens, the prefrontal cortex, and the amygdala. The VTA makes dopamine, a neurotransmitter connected with pleasurable feelings.[57] The pleasure circuit is triggered by a rewarding item or event. Once triggered, the ventral tegmental area releases dopamine into the nucleus accumbens and the prefrontal cortex. Then there is the ventral striatum that plays an important role in subjectively predicting the value, the potential reward.[58] This process produces positive memories of this pleasurable event. Recalling this positive event leads to craving a repeat of the activity. After the activity is repeated, the brain circuits involved are strengthened. More repetitions lead to even more strengthening of the circuits, making the experience more hard-wired. Some called this the dopamine-reward pathway because the release provides us with good feelings.[59]

This sets up a range of possible positive rewarding experiences. *Use* involves engagement with the rewarding stimuli, like good food, quenching our thirst, and positive social interactions, all leading to the enjoyable experience we want to repeat again. Use can flow into *misuse* when the engagement gets extended, and it is harder to stop. Not all positive rewarding stimuli are good things. Addiction becomes an issue when we compulsively engage in the rewarding stimuli and keep doing it despite the negative consequences.

The person with an addiction problem has a blend of psychological and physical cravings for the object of desire. This includes not only substances like alcohol or drugs but also behaviors used to engage in gambling, sex, ultra-processed foods, social media, and so much more.[60] Addiction can develop when the brain's reward pathway inhibitory tools, that sense we should stop now, are weakened. Technology can be highly addictive with its immediate rewarding behavior and negative consequences to our relationships. Due to how the reward pathway works, each of us can be potential addicts with access to the right conditions we find hard to resist.

	Rewarding Stimuli	Able to Set Limits, Say No	Outcome
Use	Engagement limited	Yes, can do it easily	Enjoy, do again
Misuse	Engagement extended	Yes, but is harder to do	Still can meet community standards
Addiction	Compulsive engagement	No, keep doing with negative consequences	Adverse results

It is reasonable to expect any progress in dealing with addiction involves change. A long difficult journey always starts with the first step. Stages of Behavioral Change is a tool to use when we want to intentionally work on taking the steps needed to move in a different direction. The effort starts with our Go Slow thinking making the deliberate choice to do the hard work involved. It can be very challenging to make the desired changes to our habits of mind and other behaviors quickly. Stages of Behavioral Change provides a useful guide to help us find the right road moving in the right direction to work on new ways to better manage our addiction specifically and our emotions more generally.[61]

With an addiction problem, one goes onto a long bumpy path with many ups and downs. Sometimes the positive emotion of hope is all we have in trying to find the right path moving toward a better life.[62]

Stages of Behavioral Change

Precontemplation	We do not know what we do not know. If some awareness develops, we remain in this stage if there is no intention to make any changes.
Contemplation	We recognize the challenges by becoming conscious of our incompetence in the area.
Preparation	Our intent to change is demonstrated by developing a specific plan with short-term objectives and a time to start.
Action	We are now implementing the specific plan for the change; this includes the possibility of relapse returning to the old ways of coping.
Maintenance	We have now sustained the changes for a period of time, measured in months, and intend to keep it up.
Termination	A point of unconscious competence with no interest, or temptation to return to previous status.

VI. THREAT TRIGGERS SURVIVAL MODE

Not all negative emotions are bad. Negative emotions can give us a boost when we sense that competitive spirit, stand up for our own interests, or need to prepare for the immediate challenge ahead. Generally, threat activates a biological motivation triggering freeze, fight, or flight as needed for survival. We get good at what we practice. Successfully overcoming a specific challenging threat encourages us to use the same approach next time. This, in turn, leads us into routine ways of thinking about how to deal with such challenges in the future.

When threat triggers survival mode, the body's total focus is on keeping going and living longer. The emotional state we experience at the time starts as just a fleeting moment capable of activating a cascade of unconscious, automatic biological processes inside our body. An initial startle is followed by the fog of negative emotions quickly narrowing the focus, skewing our perspective. There is only one priority in this moment.

As the threatening stimulus remains present and the situation continues, we slip into an emotional episode to keep efforts flowing. Neuroception helps, as the natural negative bias keeps our guard up. Attention filters skew incoming information in support of the current survival efforts. Muscle movements support the behaviors required. Decision-making is limited to the needs in the immediate crisis, and Quick Thinking uses memory associations appropriate to this current effort. However, under these conditions, there is no time for the deliberate efforts of Go Slow Thinking. There will be no other content considered while in this mode.

Being exposed to dangerous, threatening events has always been part of life. Predators like the saber tooth tiger in its day to well-armed terrorists today trigger the same type of biological survival efforts.[63] This can also be broadened out to include other threats such as work place issues, financial concerns, relationship conflicts, ability to meet basic needs, and other potentially damaging situations. Awareness of the flight, freeze or fight options may provide a practical approach to developing a simple plan when under imminent danger.

The natural biological defensive strategies when threat triggers survival mode include:

Flight	Mobilization	Run
Freeze	Immobilization	Hide
Fight	Mobilization	Engage

Flight is a mobilization strategy where one runs away as fast as possible to escape using the safest path possible. Sometimes this outwardly appears to be a slow walk away from a not-so-good relationship, artfully making a reason to depart and finding the exit gracefully. Flight has us moving away to avoid the threat, a solid strategy.

At other times, we may try to hide. Freeze is an immobilization strategy where one does not engage by basically shutting down. In the natural biological world, not moving is an effective strategy because predators are better at noticing moving prey. Freeze may be happening if a person is present in the room, has a stake in the proceedings, but is not participating.

When the negative emotions are connected to something important to us, and flight or freeze are not suitable options, we fight. Fight is a mobilization strategy where one directly engages the threat because of danger to life, limb, or property. We need to act with aggression, though not necessarily physical. In sports, we use this type of mobilization to prepare for the competitive game. In a political environment where one feels passionate about the cause, mobilization can lead to seeking strategies to win at whatever the cost. This can move into the collective groups sustaining a cause at the Meta Level for extended periods of time, measured in years.

There are many forms of life-threatening experiences seared into memory due to Human-Caused Dangerous Events, Unintended Human Failures or Natural Disasters.[64]

Human-Caused Dangerous Events are from military combat, mass fatality shooting, sexual assault, childhood abuse, domestic violence, intimidation, physical abuse, self-inflicted wounds, and other life-threatening situations deliberately engaged by someone.

- Warfare is brutal. One study from World War II found how every person had a breaking point despite the military wisdom of the day.[65] The research documented how after two hundred fourteen (214) aggregate days of combat duty, all soldiers were ineffective due to death, wounds, medical illness or breaking down psychologically. The key word here is ALL as in a universal finding with no exceptions! This study was before we used words like trauma.

- Adverse Childhood Experiences (ACE) Study has demonstrated how lived experiences due to maltreatment and household dysfunction translate into adult illnesses, disability, and early mortality. The ten adverse childhood experiences are physical abuse, sexual abuse, emotional abuse, physical neglect, emotional neglect, mother treated violently, substance misuse within household, household mental illness, parental separation or divorce, and incarcerated household member.[66]

Unintended Human Failure could be from injury or death due to a serious accident, or technology tragedy such as a structural collapse:

- A series of unlikely events lead to the Royal Mail Steamer Titanic hitting an iceberg on April 14, 1912, and sinking several hours later on April 15, 1912. Less than one third of the 2,224 passengers and crew survived.

- On September 12, 2008 there was a train accident near Chatsworth, California, west of Los Angeles.[67] The person with the primary duty of operating the train was also focused on texting and making phone calls. Due to not paying attention, he missed a red-light trackside signal resulting in a head-on collision with a freight train. The accident included 25 people dead, 102 admitted to hospitals and over $12 million dollars in damages.

- Part of the Surfside Condo building in Miami-Dade Florida collapsed in the early morning hours of June 24, 2021, while residents were sleeping killing 98 people.

Natural Disasters, such as fires, tornadoes, floods, earthquakes, and hurricanes as well as sudden medical issues can trigger survival mode mechanisms.

- It is not unusual to hear reports of torrential rain causing widespread flooding. At the same time, day after day of clear skies may lead to drought which triggers wild fires. These overwhelming disasters triggered by the weather, geologic, or climate events cause human death and injury as well as other damages costing billions of dollars each year.

Trauma

All of these represent adverse environmental events. We may be exposed to the dangerous life-threatening event, as in directly experiencing or witnessing the actions. Experiences external to us perceived as dangerous and life threatening represent a traumatic stressor.[68] If we survive the adverse environmental event, then a traumatic memory will be formed, which may long outlive the physical wound's healing.

We started using the word trauma because American Psychiatric Association introduced posttraumatic stress disorder (PTSD) in 1980 within their third edition of the *Diagnostic and Statistical Manual of Mental Disorders*. In 2013, the scientific inputs improved DSM-5, classifying PTSD as a new category of "Trauma- and Stressor-Related Disorders."

The concept of trauma involves events external to us perceived as dangerous and life threatening. Due to our unique individual brain, there are differences in how vulnerable one may be to the next dangerous, life-threatening event. Vicarious trauma impacts counselors, peers, medical staff, volunteers, and other workers trying to support people who survived the dangerous life-threatening event.[69] Those who sign-up to provide these trauma supports empathize with and care about other people. Listening to the terrible suffering, pain, horror, abuse, and violence experienced by many different people has its own cumulative impact over time, sometimes referred to as compassion fatigue.

As we work to provide services, efforts need to include using a Trauma-Informed approach. Trauma-Informed Care means the provider of services understands the pervasive nature of trauma. Thus, the service is organized to promote an environment of safety in support of recovery and not a service which may inadvertently re-traumatize the very person we are trying to assist.[70]

We have no control over the good and bad things that go on as part of how the world keeps turning each day. However, most of the time, we do have choices on how we respond to these events. Being scared and angry in the moment is understandable. We can be aware of choices to use in response, like wanting to seek revenge to get even or forgiving the other in order to go on with life. It may not seem so at the time when someone has done something intentionally wrong to hurt us, but there are many different choices we can make in response.

VII. RULE OF REPETITION

Consistent statements, based on actual physical evidence, scientific research, wacko-ideas, or outright lies, can become established as a social feature when repeated many times influencing our habits of mind. The rule of repetition involves saying the words and\or doing the actions in a consistent manner multiple times as needed to communicate the content. We cannot expect to say or do something once and have the target audience correctly remember the content. This is due to how attention and long-term memory operate. Thus, in practice, it means (1) knowing the core theme of a message or action to be taught, and (2) saying or doing what is involved over again steadily in order for the intended audience to retain it.

The rule of repetition involves using a process to repeat the messages a number of different times. If one is not sure of the number, start with five times. But this is not five times in one session. If the message gets overexposed, habituation sets in and the additional repetitions are ignored. This is five separate episodes of repeating the content with consistency. Once established, then repeat the content again at reasonable intervals in separate sessions by different methods. Just like so many processes in life, the method involves getting into a pattern of sustained efforts, like washing and rinsing only to repeat it again soon.

We need to notice how the rule of repetition is being used in conflicts and disputes. In our ancestor's day, to build upon group awareness, stories were told by word of mouth around the campfire. Historically, whether true or false, these stories impacted both individual and group behaviors. Negative emotional narratives, based on anger, fear, and sadness, were especially contagious. The development of the printing press provided the capacity for standardized, consistent messages. The messages could then spread by moving at the speed one could walk or as fast as a rider on a horse or a sailor at sea. The technology of the telegraph, telephone, radio, and television moved these standardized narratives even faster. Today, internet-supported social media moves these narratives around the world in seconds, as a story goes viral.

Historically and today, these narratives impact us emotionally helping to influence our perceptions of daily events as well as choices made. Our ancestors could use narratives to help develop and reinforce tribal loyalty, giving everyone a reason to work together in the fight for survival. The narrative teaching warrior mentality of no retreat, no surrender, impacted every member of the tribe. Today, the same tools are being used at the collective levels, moving at the speed of light across the internet and having widespread influence on millions of people.

For the rule of repetition to work, each step in the communication process needs to be successful. For starters, there is the ongoing competition to grab our attention. Then the message needs to pass through the attention filters in short-term memory to be initially received by the person. At the nano level, the visual, audio, chemical, and pressure signals need to be processed by the brain to gain a form of understanding. The emotional components of the message may help or hinder in paying attention, especially if functions like implicit bias effectively block messages one does not want to hear. Something important to the group perceived as a negative threat or positive opportunity will aid in getting our attention.

If the message passes the filtering actions and will be needed in more than several minutes from now, it must be consolidated into long-term memory. Associations get established as the message is formed and stored in long-term memory. At this point, the message content has been encoded in bits and pieces as a long-term memory available for recall. Then, at the moment of recall, the person needs to remember the sender's content at the right time in the right way. Unused content is forgotten.

The science of the rule of repetition involves not only emotional memory, but also association features, priming and anchoring. The association feature of memory connects the new information to an existing familiar pattern.[71] As long-term memory is consolidated, the bits and pieces of the engram pattern link the new content to other material already stored, based on appearance, sound, meaning, emotions, history, use, ideas, and so much more. If the connected association is correct, great. If there is an error in consolidation, later at recall there will be mistakes.

Priming is the first-time exposure to the content by our brain.[72] The novel, new, never seen, heard, smelled, tasted, or touched–before item is placed in our memory as a first-time experience. With this foundation established, the next experience will be linked by association to the first exposure. Some call this the mere exposure effect.[73]

An anchoring effect is connected to priming. In this case, a recommended number for an unfamiliar value is suggested by a third party before discussions start.[74] The suggested number becomes an unconscious reference point before a decision is made. We see this in shopping with the suggested retail price providing a starting point for the product's value.

We get good at what we practice. The rule of repetition uses procedural memory methods to establish, sustain, and build upon core themes. Now a long-term memory is formed and stored with a new association created. Then as our focus goes to something like an object, recall brings the associated content forward in this moment to support an opinion, pass judgment, or evaluate the item. Over time, we learn from experience to associate certain things together and then use these memories to navigate through life.

The rule of repetition is an important tool used by a variety of people to communicate the correct knowledge and skills to prepare for a job, doing the precise movements, as well as follow the ethics of the group. Educators use the methods to teach lessons to students. A coach practices the technique to prepare the team with strategies for success. Band leaders work the approach to get the musicians on the same sheet of music. Marketers try to present the finer features of their product. Writers of fictional tales may fill our minds with fun fabrications in the name of entertainment. Some of the communication is offered as suitable feedback in the form of a critique, criticism, comment, or review.

Scientific knowledge itself is neutral. It can be used for good or bad. The rule of repetition is an important engine in getting our message across to others. The fuel is the message itself. This takes us to the important difference between misinformation and disinformation.

Misinformation happens as part of the natural effort to understand and interpret our current situation when under much confusion. As part of the fog of uncertainty, we may unintentionally misinterpret the current message content, and mistakes get made as information is shared with others. Then as ideas are exchanged and suitable feedback is received, the misinformation gets sorted out with corrections made.

Misinformation is just part of living when we do not know what is happening. However, the unintentional error can be hard to correct with content moving at the speed of light around the world, compounded by our own anxiety, rumor mills, and information overload from being bombarded with news and opinion. But there was no ill intent in sharing what we know.

Disinformation is an intentional effort by deliberately organized people to take advantage of the flow of rapid communication to sow confusion, tell lies, and promote untrue rumors with the malicious intent to gain influence or money. Thus, the rule of repetition slips into the emotional side of conflict when used to deliberately organize content in order to take advantage of our feelings. The rapid communication flow has the nasty intent to manipulate people using crafty and sneaky means. A well-funded and organized group with clear assignments can engage in the systematic use of repeated messaging to spread disinformation around the world at the speed of light. Information warfare uses sophisticated techniques to drown the true facts with a flood of lies, malicious content, and false reports. Those behind the disinformation target the identified subgroups connected by their patterns of likes and dislikes. The target audience is fed carefully tailored messages, using a blend of fact and fiction. It can be used as a counter narrative to destabilize an opponent as a strategy in an active dispute. The results easily plant doubt as well as stoke existing divisions around controversial issues which triggers an immediate intense emotional reaction in the end user. The whole process makes it hard to know what to believe.

The Nazi propagandists demonstrated how to do this nationwide.[75] They used the technology of their times, radio and print, to systematically deliver their repeated talking points. They were successful for a time in brainwashing their audience to believe lies, half-truths, and deceits. Millions of people died during World War II due to the impact of the Nazi propaganda.

The rule of repetition is the core action needed to effectively use propaganda. It involves repeated expressions in a steady pattern over an extended time to simplify complex concerns, creating a bias in favor of one side. The fuel of the message changes the unresolved conflict to a small protest growing into a huge, catastrophic dispute on to full warfare as negative emotions go off the track. This is an important tool when the stakes involve winner takes all and the loser goes away with nothing.

These deliberate efforts make the rule of repetition into a tool of power. Power is basically the capacity to have influence over others to achieve desired results.[76] The sources of power come from the combination of leadership, organization, and money. Money allows property like land, buildings, or equipment to be purchased as well as the skills of people to use these resources. In combination, these sources give the leaders access to the tools of power in order to influence the emotions of people involved in these issues. Hard power is a visible tool using force or favor to gain compliance. Soft power uses more implicit approaches to attract and persuade people to join. Was hard power used to compel people to join in? Was soft power used to inspire people to arrive on time and voluntarily participate?

The rule of repetition works as an effective tool of power in the politics of fear and favor. Its use can range from frivolous to substantial. As a threat triggers survival mode, we project strength as we keep our guard up. Anything contrary to our group's point of view is opposed as we do engage in this fight using the repeated message strategy. Meanwhile the reward pathway is reinforced by favors responding to questions of what is in it for me, or why should I care? Also, success with these methods encourages the continued use of the strategy.

All of this comes into consideration as we work to resolve disputes. So, it is important to notice when others are using a rule of repetition method. Understanding the rule of repetition helps us to recognize when a salesperson, politician, or another with an agenda is trying to emotionally exploit and manipulate us. The simple goal is to influence the decision-making of their target audience. Emotions are directly connected by giving meaning in the moment, by the filtering action of attention, by forming and storing in long-term memory, and by the triggers in recall of the memory, leading to the choices made in the moment.

The rule of repetition has the capacity to not only teach positive lessons, but also to get people all riled up, agitated, upset, irritated, fueling anger. Some individuals constantly tell blatant lies, followed by denial when confronted with clear evidence of the deception. Talk is cheap. So, when the words do not match up with the actions, it is a time to ask ourselves questions. Focus and pay attention to what other people are saying, listening for the emotions behind their words and behaviors. This is the job of Go Slow Thinking. If the words spoken contradict the person's actions, facial expressions, or related physical signals, it may indicate someone is lying. The person may say one thing and do another, often maintaining a double standard.

Misinformation and deliberate disinformation campaigns are created, spreading around the world at the speed of light.[77] Distractions only add to the confusion. This false, distorted material appears just a real as the factual content, making it very hard for people to tell the difference. Then when the facts involved run contrary to what we want to believe, it becomes easy to proclaim a problem. As a result, factual information becomes a choice to believe, discount, or ignore it.

Freedom of Speech and Fact Checking

The whole information-sharing process can set up a clash of individual rights in the free exchange of ideas. Under the Constitution of the United States, the Bill of Rights includes the freedom of speech.[78] From this freedom, a marketplace for ideas has emerged where we have the right to say whatever we want, listen to what we want, ignore certain things, and deny the reality of certain facts. As with many things in life, there are limits to this freedom of speech. With each right comes a responsibility. Acting responsibly includes making the choices on content used as well as being held accountable for one's conduct. When false claims cause serious harm, the defamation may lead to lawsuits, especially when the one violated can show the written word was false and who was responsible.[79]

More often we limit ourselves when hearing such content. When there are concerns raised, we need to ask who stands to benefit from it. Then we ask questions and work to follow the money trails to see who benefits. Looking at the underlying facts can make the whole claim seem obscure, murky, unclear, and incomprehensible.

Concepts of Evidence

When we want to sort out the truth from the fiction by increasing the objective features as well as tamping down the subjective element, consider the concepts around rules of evidence. Both legal and scientific fields use Go Slow Thinking to collect evidence and build a case for or against what is observed. Science deliberately collects objective measures to reveal the realities of the world with the general intent to improve our understanding. Both science and legal evidence standards range from a well-informed guess to solid, factual evidence.

The concept of evidence sets up a range of options when fact checking a message.[80]

- The legal concept of *probable cause* is similar to a scientific hypothesis where there is enough evidence to establish a basic suspicion. In science, basic suspicion leads to a research study. In law, the basic suspicion of the crime leads to someone being charged.

- In legal circles, *preponderance of evidence* would be like when in science, a research process including peer review yields interesting results, leaving one with a sense that this content has greater credibility to be true.

- Legally obtained material considered to be *clear and convincing evidence* is like when in science, there is additional reported success with the same process, leaving observers with a firm sense this material is true.

- Legally obtained evidence meeting the high standard of being totally convincing that the content is true is called beyond a *reasonable doubt.* After many times of success, the scientific method produces a constellation of results supported by robust data sets from many researchers supporting the ongoing findings as totally convincing.

DESCRIPTION	Legal Concept	Scientific Method
BASIC SUSPICION	Probable Cause	Scientific Hypothesis
GREATER CREDIBILITY AS TRUE	Preponderance of Evidence	One research process with interesting results
FIRM BELIEF THIS IS TRUE	Clear and Convincing Evidence	Additional success with the same process
TOTALLY CONVINCING	Beyond a Reasonable Doubt	After many times of success, the robust data sets support the previous findings.

Trust, But Verify

The whole thing can be a vicious cycle. Our attention filters what we experience. Our memory system not only forms, stores, and recalls bits and pieces of those experiences but also forgets certain content. In turn, these memory functions influence our perception of reality and beliefs. It can be hard to let our guard down and trust unfamiliar sources or those with a questionable track record.

When negotiating with Mikhail Gorbachev, the leader of the former Soviet Union, President Ronald Reagan said, "Trust, But Verify." Verification is the deliberate work of going to the original source to confirm the reported content. A good practice is to seek two independent sources to confirm the content. The sources cannot be from the people making the claim or any of their supporting organizations. So, this idea of trusting and verifying requires deliberate efforts in seeking, reviewing, auditing, and evaluating the reported content.

Senator Daniel Patrick Moynihan is remembered for noting how everyone is entitled to their own opinion, but not their own facts.[81] Trusting but verifying, as well as sorting opinion from fact takes time and the deliberate effort of our Go Slow Thinking. "You can fool all the people some of the time and some of the people all the time, but you cannot fool all the people all the time" is a popular saying.[82] When "trust me and not the other guy" is a theme, we may be picking up someone using a rule of repetition approach, working to turn lies into "truth."

Detecting the cheating requires discovering the facts and doing something with them. These are good habits of mind to develop in a world filled with endless criticism driven by the us-versus-them dynamic. However, the work takes time and effort. As a communication tool, the rule of repetition can feel like slow-motion change using a thousand little messages. The cumulative effects interact and add up to give the practitioner positive results in shaping our everyday life. When addressing conflicts and disputes, we need to notice how the rule of repetition is being used.

VIII. CONFLICT DRIVEN BY EMOTIONS

A well-known Benjamin Franklin expression is "Certainty? In this world nothing is certain but death and taxes."[83] The dynamics of human life includes the reality of change. Nothing stays the same. Sometimes the changes are fast while others take a very long time. At times, it feels like progress moving forward while at other moments, it seems like we are losing ground. Some people promote certain changes while others resist. It never stops.

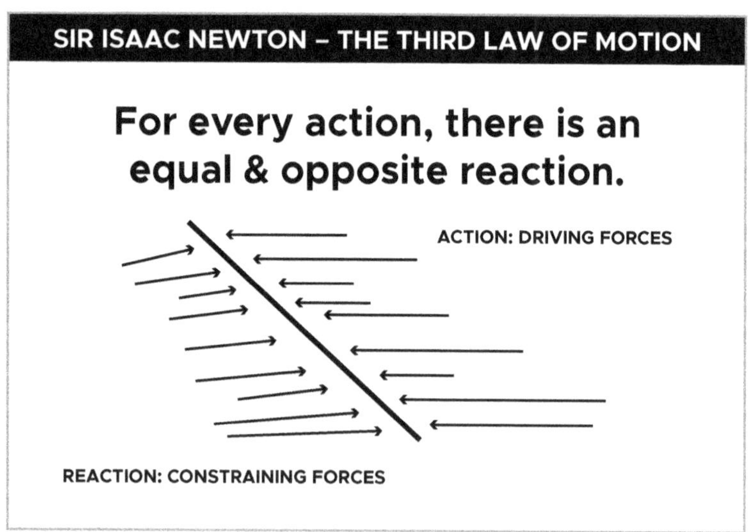

Sir Isaac Newton's Third Law of Motion says for every action, there is an equal and opposite reaction. With the ebb and flow of this ever-changing world, life sets up plenty of uncertainty and conflict. Our human brain has the capacity to adapt to an ever-changing world. The brain is also where all conflict begins and ends, driven by our emotions giving meaning in the moment. Over a lifetime, our human brain is constantly changing as we grow from newborns into toddlers on to school-age children, adolescents, adults, and seniors. Each age has its own needs and demands. What is important at age three is not an issue at age thirty-three. However, some things about the human body never change. The universal biological basics of life, driven by our shared 99.9 percent genetics, mean we all physically need food, water, and shelter for a safe place to sleep. How those basic human needs are addressed does change.

The biological reality of human emotions is the central driving force underneath conflicts. So, it is important to understand how the biological reality of emotions and thinking are connected in order to better cope with all the forms of conflict.

- Conflict is simply part of living life. There are a wide variety of reasons for conflicts forming in our minds, as we sense something is wrong. It may be based on live events, something recalled, or imagined.

- The word *conflict* implies struggle, ill will, and hostility. It can range from some mild discomfort inside our mind, to a sense of differences we have with another person, to active competition between groups using offensive and defensive tactics as well as an enduring battle of ideas.

- The conflict becomes a *dispute* when one side makes a direct claim against another and is now willing to take action.[84]

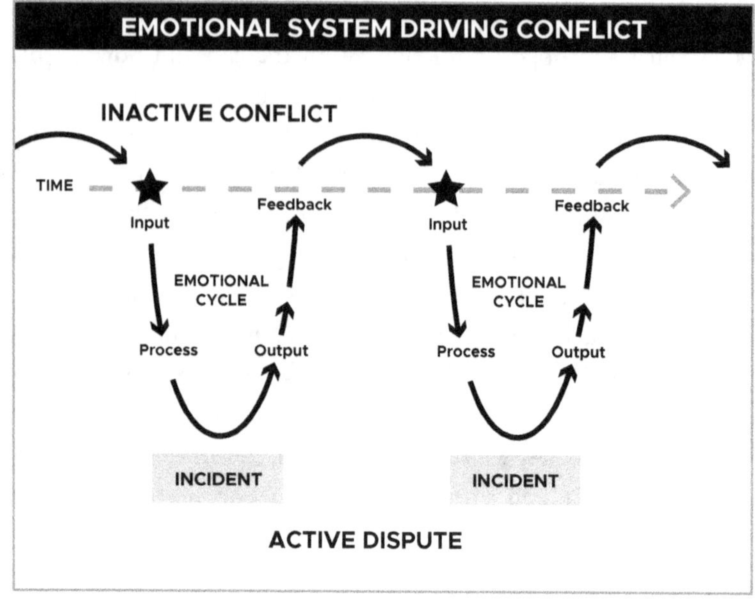

Emotional System Driving Conflict

We can take our knowledge of the emotional system and the challenges of conflict to notice a predictable cycle. Each step of the cycle provides a chance to stop the conflict and possibly end the dispute.

- **INACTIVE CONFLICT**—On the surface it appears all is well as the unresolved conflict lays dormant under the surface for extended periods of time.

- **INPUT**—The right cue, signal, or trigger strikes a spark setting off an emotional state to start a cycle of conflict. The open state of hostility is now an active dispute.

- **PROCESS**—The active dispute is driven by emotional states quickly fueling the process, and the clash is fully engaged. The emotional state may be experienced as fear, anger, sadness, contempt, concern, wonder, hope, and more. The ongoing nature of the dispute becomes an emotional episode transiting one emotional state into another. The whole dispute casts a certain mood remaining consistent for extended time periods and coloring our perceptions of the conflict.

- **OUTPUT**—These are the focus, thoughts, judgments, muscle movements, and memories recalled used to deal with the conflict.

→ Attention's filtering action narrows the focus in the moment.

→ Habits of mind (Quick Thinking and Go Slow Thinking) influence the flow of thoughts.

→ Muscle movements used for facial expressions, posture, and tone of voice as well as the current behaviors engaged to address the dispute.

→ Long-Term memory brings the past to be part of our present moment as the recall of emotional events influences attention, thinking, and our current actions. Long-term memory captured and retained the important experiences in our life for potential use in the future, stored as bits and pieces. Now, in this time of active dispute, selected bits and pieces of useful memory are recalled to support our efforts.

→ Neuroception is like a protective shell in place to help safeguard our self-interests. The guard is up as part of the natural, continuous, unconscious monitoring for dangers and threats. Inside the room until a sense of safety and trust gets established, flight, freeze or fight gestures are used to deal with the moment.[85]

  • Flight—Refuse to negotiate by just walking away or not showing up.

  • Freeze—Shut down; one is present but does not participate. This may be due to being overwhelmed, tired of the battle, unable to mentally process the proceedings, avoiding the sensitive topic, or just using a passive tactic as a strategy.

  • Fight—attack every proposal from "them," be very challenging with no compromise, demanding "it is my way or the highway." It may also involve one who is naïve about what is a realistic outcome under these conditions.

→ Judgments include opinions expressed and decisions made reflecting our preferences. Conflicting loyalties can force choices in order to stay connected with our preferred group. Sustaining the sense of belonging can be a powerful influence.

- **FEEDBACK**—The aftermath is the time period following the actions where the consequences become known and feedback is given. It can be a time to pause, considering what went right, possible areas of improvement, and next steps.

- **INACTIVE TIME**—Everything settles down for now, however, tensions may still be bubbling away under the surface. Then another trigger event starts the next cycle (Incident 2) and followed by more inactive time, followed by Incident 3, and so on.

Go Slow Thinking needs to be used if we want to break the cycle of conflict. The initial efforts involve holding our response while weighing the options. Outwardly, we are friendly and cordial. We acknowledge the other point of view. However, for now, we keep our available options to ourselves, restraining comments, giving nothing away.

Meanwhile, inside our mind, we consider these options. This is the time to think through what is important under these conditions and what can be ignored or dismissed.

A wheel, such as used for a wagon or bicycle is constructed using a central hub, a number of spokes, and an outside rim. The outer edge of the rim is connected to the core hub by the spokes. In order for the wheel to remain a functioning structure, a certain number of spokes need to be present.

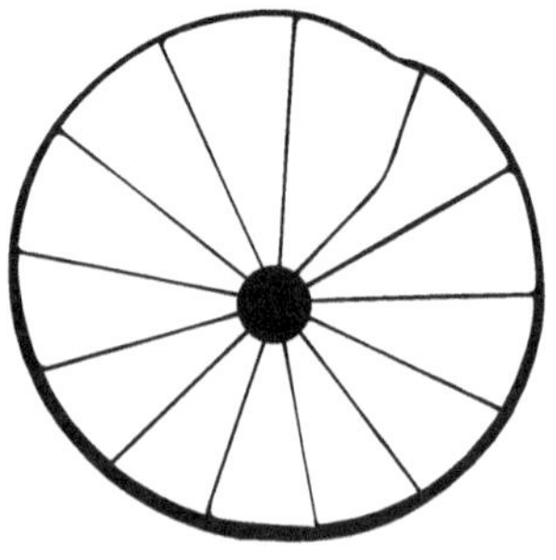

In entering such a discussion to resolve a dispute, we need a clear picture of our own core interests, values, and limits. We can think of the hub as the core of what we need to make the efforts worthwhile. The rim can be considered the outside limits to which we are willing to go. Then the question becomes, how many spokes can be given up and not lose the functional integrity of the overall wheel? It is possible through the efforts to reach an agreement, we gain a spoke or two, strengthening our wheel. When ready, we can use Go Slow Thinking to choose our response in a calm and proportionate way. Then the other party to the dispute has the option to match our response in a calm, respectful way or simply trigger another cycle of conflict.

All of this can be overwhelming. A good rule of thumb is to try a tit-for-tat strategy where we match their approach.[86] If starting the discussion, we show a cooperative approach as our first move. Then watch what the other side does. If they respond with an adversarial approach, we can make an adjustment, matching their tactics. If they keep showing a closed-minded, fault-finding, adversarial approach, these negotiations are not likely to work out, so we walk away.

However, if they show a cooperative approach, match that. Over the long term, the cooperative approach is a solid strategy for the best possible mutual gain. But each situation is different, and we need to be aware of the option to make a deliberate choice fitting the moment. Generally, one may expect both sides to get locked into a steady pattern of cooperation or hostility. So that first move is important.

IX. LEVELS OF ORGANIZED HUMAN ACTIVITY

For teaching purposes, four levels of organized human activity can be described, setting up the emotional side of conflict.[87]

Individual Levels

NANO LEVEL is intrapersonal, playing out inside our mind and body including self-talk. The term is based on the reality involving how space between two neurons is measured in nanometers. Nano level functions are invisible, tiny, microscopic events measured in milliseconds and microns, building into patterns across our body, moving from the unconscious into the conscious. The time frame plays out in seconds, minutes, hours, days, or longer.

Nano habits of mind are routine ways of thinking inside our brain, developed over time to help us cope with living. Inside our mind at the nano level, we are blind and deaf to our own good, neutral, and toxic implicit bias while it actively influences our mental inclinations and worldview. These can be automatic ways of thinking, using well-practiced, long-standing scripts from our Quick Thinking. When encountering something novel and new, our Go Slow Thinking can be used to consciously and deliberately consider what is happening. From these nano habits of mind we interpret, classify, or categorize events, ideas, and information (good/bad; like/

dislike) in order to understand the experience shaping our reality. Some of these nano habits of mind are very useful in helping us cope. Others may actually be harmful by skewing our perceptions, thus distorting the true situation.

MICRO LEVEL is between individuals. It is the place for the interpersonal, small intragroup relationships we have at home, work, and other social settings. The time frame plays out in minutes, hours, days, weeks, or longer. Between individuals at the micro level, the reality of culture defines what is a threat or opportunity, ethics of right and wrong, and what it means to be us within this group versus one of them.

Micro habits of mind involve two unique brains forming one unique relationship. It is a set of common agreements between two individuals working to form one team, such as in a household or work setting. Life has a way of bringing two people together. However, once found, keeping things moving in the right direction cannot be left to chance. Even if the relationship feels like a natural fit, keeping it going requires careful nurturing as well as a commitment of time and effort. The relationship needs to be a priority for both parties in order to strike the right balance and manage the inevitable complications.

Collective Levels

MACRO LEVEL involves large groups working within the same culture. It is the place for intergroup relationships connected together by a shared culture, using the same jargon. Within the same culture at the macro level the dynamic leader serves as a role model for other members of the group, teaching the jargon, sense of right and wrong, how to manage the ongoing relationships, and more. The time frame plays out in months, years, or longer.

The macro habits of mind forming the social contract helps the group to organize and guide what it means to be part of this tribe.

In between the "universal all" and "unique one" are things groups of people share in common, such as language, units of measurement, as well as values. Within the memory system, our attention filters include a partisan lens to very quickly sort out who is positive (member of our tribe), negative (oh, one of them) or can just be ignored. We commonly move toward other people sharing common interests and away from those in conflict with our ideas. Groupthink involves a team of people interested in a cohesive mode of thinking with a desire for consensus.[88]

META LEVEL is where the decisions get made to organize overall society. What cannot be managed at the macro level gets sorted out at the meta level. This is where the work of addressing competing interests happens as the political will of the top officials shape social, public, and institutional reality for the whole of the people. The top officials develop bylaws to establish forums for debate and decision-making like legislative bodies, judicial courts, as well as executive leaders. The interactions involve work between different nations, cultures, industries, and other groups. The time frames are very long, playing out over years, decades, or more.

The top officials shape the social structures for the whole of the people by using these meta level forums for debate and decision-making. The winners of the struggle to control these forums engage in the work needed to address the various competing interests. They set the course to follow, then use implicit and explicit methods to implement the choices made. Meta habits of mind are the routine ways of thinking needed by leaders to manage their duties as set by the governing authority to organize overall society. Through election, appointment, or other means, macro leaders are selected to serve at the meta level.

The meta leaders are people bringing to the table all of their human strengths, weaknesses, and implicit bias. They cannot be all things to all people. Priorities need to be set. Hard choices get made, based on the emotional sway of the people involved. In the process, lines get drawn, looking for the balance between the various interests and the greater good.

Geologic and Astrophysics

Meta level time frames can feel overwhelming. After all, they do stretch well beyond a single human lifetime. However, the meta level of organized human activity pales when placed in relationship to geologic and astrophysics time.

Geologic time covers the four-point-five (4.5) billion years of Planet Earth's history. This takes in how mountains were formed, the movement of Earth's crust called plate tectonics, and much more. The entire human history of settlement and agriculture is only measured in thousands of years, a mere speck of time relative to Planet Earth.[89]

The science of physics is used to study how galaxies, stars, planets, and more form in order to reveal how the universe works. While hard to grasp, astrophysics works to measure the vastness of space. One astrophysics unit of time measure is light-years as in the distance light energy travels in one year. Astrophysics time units involve millions and billions of years.

The Voyager 1 and 2 spacecraft missions may help to demonstrate these distances. Launched in 1977, the two Voyager spacecrafts were both designed to reach the edge of our solar system. The interstellar missions of Voyager 1 and Voyager 2 was achieved by the spacecrafts moving at a velocity of fifteen (15) kilometers per second or about 35,000 miles per hour. It took thirty-five years of space travel for Voyager 1 and forty-one years for Voyager 2 to leave our Sun's influence and enter interstellar space. In 1996 when the spacecraft was over four (4) billion miles away, Voyager 1 took an image of Planet Earth.[90] Within the incomprehensibly vast distances in space, that image of Earth looks like nothing more than a pale blue dot.

X. THE IMPLICIT BIAS PHENOMENON AND SYSTEMS OF PREJUDICE

The outcomes from the emotional system, including the implicit bias phenomenon, influence not only when conflict slips into dispute, but also how human activity is organized between individuals and collectively. The long arch of history from around the world provides plenty of examples on how the implicit and explicit features of emotions flowing from nano habits of mind up to the meta level shape our collective community.

At the collective level, a system of prejudice may be established by a dominant group, placing restrictions upon a designated set of people based on selected features like skin color, gender, age, ancestry, religious preferences, geographic home, political leanings, economic status, or any other indicator. The system of prejudice is sustained by a chain of bias transmission, using a seamless blend of nano implicit and explicit features flowing up to meta level decision-making.

It is important to acknowledge and be honest about our shared human past. The pattern of meta level actions to establish, sustain, and modify systems of prejudice is common. Racism, sexism and other forms of intolerance get deeply rooted across wide aspects of society, emerging as implicit and explicit features. Over the course of time, we find rigid gender roles in various cultures. The laws authorizing slavery and Jim Crow segregation did not happen in a vacuum.

Consider Nazi ideology in Germany (1933 to 1945) and South Africa's apartheid (1948 to 1994). Think about the challenging relationships between Protestants and Catholics in Northern Ireland as well a Sunni and Shia in the Arab world.[91]

History demonstrates the meta level time frames used to establish systems of prejudice. Then our ancestors' explicit actions in the past supported or reduced these systems of prejudice, impacting life today. The winners of the meta level struggles approve the rules, write the history, and teach the lessons. Explicit links in a chain of bias transmission can sustain systems of prejudice for extended periods of time. Even when the explicit authorization is removed, the legacy gets so deeply rooted, the implicit effects continue.

At an individual level, perception of an ongoing threat forces one to live with a sustained need to keep the guard up. This sense of ongoing threat becomes part of self-talk inside the privacy of our mind.[92] Experienced as nano habits of mind, our brain constructs these scripts, heuristics, narratives, and automatic thoughts to help manage the daily flows of living. The inner speech inside our head provides a steady stream of connected thoughts, building into sequences that are linked together.

As one person's line of thinking is shared with another, *a chain of bias transmission* is started.

A chain is a series of connected components setting up a sequence of items, as with a set of connected metal rings, a supply line, or the spread of infection. A chain of bias transmission moves a sequence of thoughts, using the implicit and explicit features of our emotions, from one person to the next. This may be good or bad depending on one's perspective. A chain of bias transmission can sustain both good ideas like the ethics of right and wrong as well as negatively motivated actions like prejudice.

Within a culture, there are unwritten rules guiding our choices, followed with no discussion and without question. We know these as conventional wisdom, the old adage, a rule of thumb, and truisms, which are simply understood by the group members. They become a concern as vectors of the system of prejudice, flowing as part of a chain of bias transmission when their core myths just refuse to die even in light of strong evidence to the contrary.

Macro leadership sets up a chain of bias transmission by using a pattern of frequently repeated messages that impact how people think about the questions of the day.[93] The macro level messages and the meta level disputes influence the thinking of everyone involved. The reality of culture is sustained by a chain of bias transmission over time via ways like music, art, literature, business practices, schools, and religious services. The explicit and implicit lessons are also taught from parent to child, teacher to student, manager to worker, friend to friend, and peer to peer, becoming the new links in the chain. The message can also flow by a deliberate, sustained, and entrenched disinformation campaign. Habits of mind are hard to change as the messages impact how people think about a given topic.

Hostility and Cooperation

As we consider this history and unpack the science of emotions, there are several factors to keep in mind. With the ever-advancing technology changing how we live and the various other challenges faced today, it is easy to forget the basic fact that our human brain and emotional system are the same as our ancestors. We have the same universal body features as "The Iceman" who was alive over 5,000 years ago. He was found in 1991, buried within an Italian Alps glacier remarkably well-preserved.[94] He shared the same universal body features uniting all of us today. He also needed his emotional system to survive while dealing with threats and opportunities of his time. However, due to one's lived experiences, there are differences which give him and each of us a unique brain.

Today we live in a very complex, constantly changing, and diverse world with billions of people. Connecting all these people are emotions giving meaning in the moment, flowing from the nano level inside our mind, to the micro between individuals. This in turn influences important social features helping to shape our collective community flowing from the macro level of large groups within the same culture into the meta level, organizing overall society. These connections help to link everyone, which in turn can promote both hostility and cooperation. As an approach to life, we need to keep our guard up in a crowded, adversarial, competitive world where some are always looking for the partisan advantage, in a zero-sum game, winner takes all.

At the same time, our survival also depends upon knowing whom we can trust leading to cooperation. Life is hard. It can be very cruel and unforgiving in dealing with threats and opportunities. When connected to a group, we improve our odds for survival. We are also influenced by the people around us in the group. Our brain's use of emotional tools helps us to identify who to cooperate with and trust or help us avoid those who may cheat us. More often, life presents us with a mix of competition and cooperation. For example, we need to compete to be selected to join the group and then cooperate to help the team succeed.

At the collective level, losing the battle does not mean the final results need to be accepted. If one has the leadership, resources, and organization, it is easy to reframe events to have new points of view regarding the issues involved.[95] It is a choice for the leader at the macro level to set the tone for the group going forward. The choice may be to cooperate with others or be adversarial, looking for the gotcha moments.

Where Do We Go with Next Steps?

A system [input, process, output, feedback] is perfectly configured to produce the current results. So, if one is satisfied with the status quo, fine. If we want different results, the system needs to be changed. At every level of organized human activity, there is opportunity to try to change a chain of bias transmission if the suitable feedback is considered. In developing a strategy to break a chain of bias transmission, one needs to consider all the components of the system of prejudice feeding into the implicit and explicit features. The well-intended quick fix, featuring the blame game, may cause more problems than it solves. A chain of bias transmission is sustained across the four levels of organized human activity (nano, micro, macro, meta) supported by the emotional system. Without insight on how all these features influence each other, finding the sustainable changes is impossible, dooming one to reexperience the same problems over and over again.

History shows those gaining from the system of prejudice will resist any changes. Sir Isaac Newton's Third Law of Motion, every action creates an equal and opposite reaction, is in play. It is hard to dismantle the system of prejudice while the dominant group works to retain the status quo for years at a time. The starting point is to acknowledge the effects of the deeply entrenched bias at both the individual and collective levels. To make system changes takes deliberate repeated efforts over long periods of time with the expectation of serious resistance from a well-organized "them." Without such an effort of sustained actions to make changes, the existing systems of prejudice do not budge.

XI. IMPROVE RELATIONSHIPS AND REDUCE CONFLICT

We can work on improving relationships and reducing conflict by putting science into action. In this case, the action represents understanding the biology of human emotions' influence on our attention, thinking, behaviors, and decisions. Being familiar with the basic science of how emotions function enables us to use this knowledge to better our lives. Here are twelve strategies to apply the science of human emotions in order to improve relationships and reduce conflicts.

1. Interests, Issues, Positions

We can use identifying the interests, issues, and positions as a way to organize our thinking about the conflict. A good first step is to recognize what is driving the conflict and dispute. Underlying all the concerns are our emotions, supplying the meaning and fueling the urge to address the basic challenges but can become something much greater. Use Go Slow Thinking to consider what is happening by making the conscious effort to clarify what is driving the disagreements.[96]

> → *Interests* are the needs we want to satisfy, such as food; clothing; and shelter; the desire to be treated with respect, be understood, have control over our life, privacy, safety,

financial security, or trust. Interests may be based on princi-ples, beliefs, and values. It might involve control over mate-rial goods, reputation, job responsibilities, payment due for services delivered, liability for damaged property, parenting time with the children, and much more.

→ *Issues* are the problems or topics we consider basic to the dispute based on the interests.

→ *Positions* are specific proposals or demands, framed as solu-tions to resolve these issues.

Interests - the needs we want to satisfy.

Issues - the topics considered basic to the dispute based on the interests.

Positions - Specific proposals, framed as solutions, to resolve these issues.

2. Active Listening.

One essential task in resolving a dispute is communication because without it, we are dead in the water. Paying attention to a speaker is a vital part of a total communication process. The speaker's words are encoded and sent out as a message. The message is received and decoded by the active listener. One job of the active listener is to make the effort to pick up on the emotions of the speaker, not only in the content of the message, but also as displayed by tone of voice, facial expressions, and posture. This is made harder because part of the decoding process is influenced in the moment by the active listener's own attention filters reflecting beliefs, concerns, and current frame of mind. Hearing is the reality of sound waves hitting the ear drums and then biologically processed by the brain. Active listening uses hearing as the single focus and primary duty at this moment in this space as part of a communication cycle.

Successful communication requires both the speaker and listener to be fully engaged in this process. Active listening is a skill developed over time with practice. We get good at what we practice.

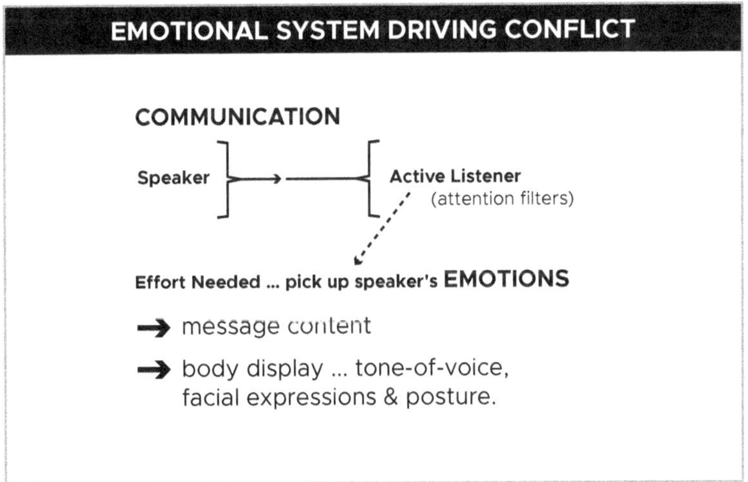

3. The Art of the Question

An important skill in managing our emotions is learning how to ask questions. A question is a sentence used to prompt a response or gather information as part of the communication process. It includes learning how to use the four rights: (1) asking the right question (2) in the right way (3) at the right time (4) to the right person (or source). The art of the question involves developing good habits of mind, using both Quick and Go Slow Thinking. For example, fact checking starts with more questions than answers.

THE ART OF THE QUESTION

Is asking
- ✔ the right question
- ✔ in the right way
- ✔ at the right time
- ✔ to the right person (or source).

TWO STEPS:
ASKING, THEN LISTENING

Most of the responsibility for the art of the question falls to our Go Slow Thinking, making the deliberate choices on the approach, content covered, phrasing, and timing. We can show interest by asking questions and then being silent while actively listening to the reply. We need to be in the moment giving our full attention to the person. Like inhaling and exhaling are the phases in breathing, so too, the art of the question requires two steps of asking and listening. It does us no good to prepare just the right question only to ignore the answer.

4. Acknowledge the Heat

A burning fire requires fuel, oxygen, and heat. To put out the fire, we need to remove one element. In conflict, a simple acknowledgment can work like removing some heat from the fire. Most of the time, we can start the cooling-down process by acknowledging the person's emotions ("It looks like you are upset") as well as their point of view by actively listening and then repeating back some key points. Acknowledgment does not mean agreement with the other's point of view. Acknowledgment does mean taking the time to deliberately pay attention to the other person and then let them know we were listening. Acknowledgment takes the time and effort of Go Slow Thinking.

5. Concept of Consensus

When working to reach an agreement to resolve a dispute, it is important to recognize when we *can live with* a solution but not like it. Emotions provide the biological processes used to shape the meaning out of the options faced to resolve the dispute ranging from love this, support it, like it, <u>can live with it</u>, to <u>just cannot live with it</u>, dislike it, oppose it, have a strong aversion to it, or are just plain disgusted by it. The default setting is being constantly on alert with guard up, as emotions start with the task of looking out for our own self-interest.

At the same time, we have the social need to establish and sustain relationships with other people. At the start of discussion, we can focus upon what is at stake for both by attending to each party's interests. The goal is to change the dispute from a competition of us-versus-them toward a more cooperative us-versus-the problem. Now the focus becomes a shared problem for both sides, making consensus possible.

An agreement will include some concessions and compromises needed to find the workable solution we may not like but can live with. Honoring agreements by consistency between what we say and do builds trust. Also, in building consensus, make sure each participant can point to some kind of "win" within the package of agreements. This is all part of developing sustainable relationships. As the old saying goes, united we stand, divided we fall.

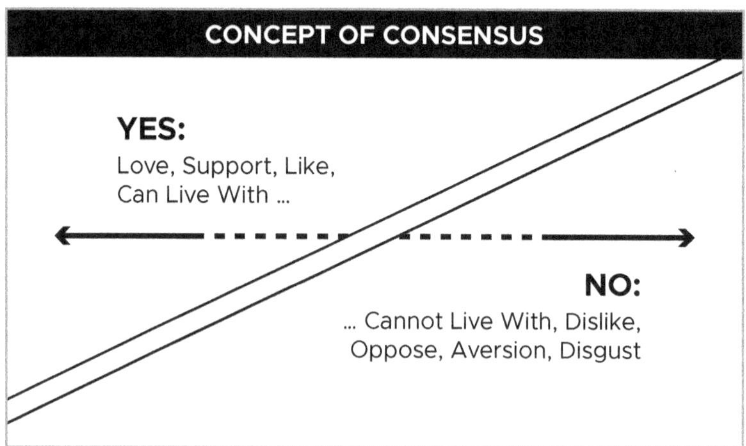

6. Learn to Set Boundaries

Our emotions give positive and negative meaning to everything we deal with in life. Without the emotional input, everything is neutral, and we just would not care. Thus, setting boundaries is about personal choice at the nano level, as influenced by our emotional system, indicating what we like (positive emotion) and dislike (negative emotion). At the core of setting boundaries is the willingness to say no. This is learning how to draw the line in the sand, setting a clear limit to establish the boundary. Part of setting boundaries is knowing our own limits. It is easy to overestimate our ability to properly engage in this dispute. This is the moment to recognize when the situation is beyond what we can manage. We now need to seek additional help or just let it go.

7. Eustress and Distress

Not all negative emotions are bad. Our perceptions sensing threatening events can be useful by helping us to attend to what is really important in the moment. Stress is part of the daily demands of living. Learning how manage these emotions can help us get through very challenging times.

Dr. Hans Selye (1907–1982) was an endocrinologist who introduced the concept of stress within a biological context, based on extensive research.[97] He rose to world-wide recognition due to his work laying the foundation needed for others to further research stress. From this pioneering work, we talk of stress management. We do not talk of stress elimination. Stress is the natural biological response to life experiences.[98] With acute stress, our body ramps up the resources needed to immediately address the threat. It can take around ninety minutes for our metabolism to settle back down after the freeze, fight, or flight response is over.[99]

Challenge stress or eustress is there to enhance our performance with just the right stimulation to arouse us into action. It is what we use to cope with the high demands of the moment with some helpful resources to handle the circumstances. A certain amount of stress can motivate us to get up and face the challenges of the day. This is what we use to prepare for and perform the challenging work, doing what needs to be done. Eustress enables us to meet the physical demands of the challenge and have the will, the mental focus, and the determination to attend to the situation. This beneficial stress motivates us to practice and prepare for the task at hand, increasing our odds of success.

Then there is distress, or threat stress, where the intensity of the strain places a variety of negative forces on our body. We can feel overwhelmed from the forces pulling and pushing us while trying to handle a situation. Over time, our body can become accustomed to the distress, and we get numb to the sources. When the sources of the threat remain constantly present, we must continuously keep our guard up against the ongoing challenges. These sources include relationship issues, job pressures, money difficulties, transportation challenges, health problems, inadequate diet, media overload, vicarious trauma, and poor sleep. Burnout can occur leaving us feeling physically exhausted and emotionally drained due to this prolonged stress.

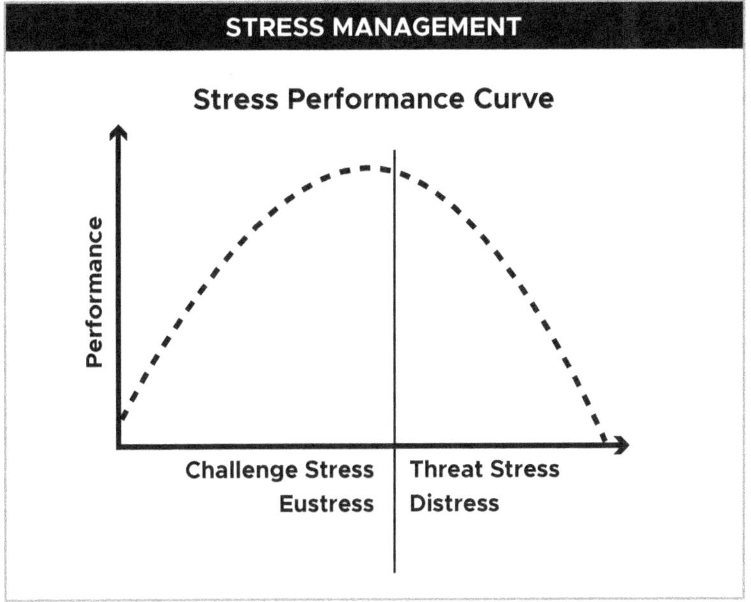

Stress does need to be managed and is a part of the way we properly maintain our body. If not, we will experience problems due to the basic wear and tear on our body, which arise from the constant demands. The stress response represents a state of physical and/or psychological arousal powerful enough to get us to stop what we are doing right now to focus upon the threat. Under threat conditions, our body responds by using available resources which may throw our ability to maintain homeostasis temporarily out of balance. After the threat has been addressed, the body needs to rest and recover in order to restore homeostasis.

Strategies to cope with stress involve finding the right balance between negative and positive emotions. As discussed above, not all negative emotions are bad. The other side of this is also true, that not all positive emotions are good. Consider how the brain's reward circuit gets hijacked, pursuing opportunities involved with the vicious cycles of addiction.[100]

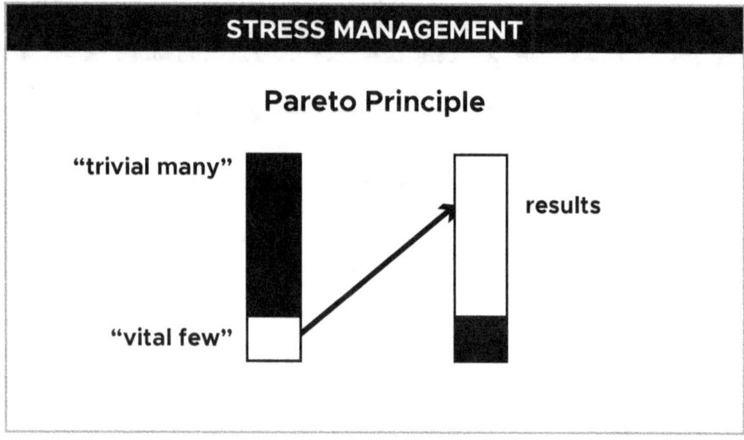

One strategy to cope with stress is time management, a form of setting boundaries. Time as a reality is based on astronomical observations, specifically the rotation of the Earth (twenty-four hours) as well as the Earth's orbit around the Sun (one year). From there, human culture has emerged to help organize a way for all of us to coordinate our activities with measurements of duration as in how long specific intervals last, like a minute, hour, day, and so on. So, how do we spend our time? As a time management tool, the Pareto Principle establishes the idea of identifying the vital few activities which yield numerous results in contrast to the trivial many actions which yield less.[101] The pareto principle is not a science-based ratio, but it is often referred to as the 80–20 rule with the idea of separating the vital few activities from the trivial many. Here we can consider our individual capacity to control how time is spent. How do we spend our time each day? A certain amount of time must be devoted to basic living activities like eating, sleeping, hygiene, and related tasks. Time needs to be used for earning money, maintaining the household, and transportation. There are those other areas we enjoy but can burn up a lot of time. Then there are the questions on how do we actually focus our attention. Do we have some self-control difficulties? The idea here is to establish a set of clear priorities separated between the *vital few*, essential tasks versus the *trivial many* activities.

8. Forgiveness

This is a process to release all the bitterness and anger retained from the conflict.[102] Forgiveness is a voluntary choice to no longer carry a burden into the future by closing the book on the episode. Forgiveness does not mean we forget what happened. Fool me once, shame on you; Fool me twice, shame on me. It is a unique choice for the individual playing out over time, not happening all at once.

- Forgiveness does not have to include an apology from the offender acknowledging the error with some kind of expression of remorse.

- In groups, forgiveness is the tool to help restore members back into a good standing, helping to sustain the needed relationships. Since we know mistakes will happen and no one is perfect, forgiveness provides approaches to correct the errors made. This, in turn, helps the group to function better for the long run.

- We can also forgive ourselves for any past errors. Nobody is perfect. We all have flaws and make mistakes. Acknowledge the heat, this time inside us. Try not to let those past mistakes define our lives. Forgiveness is something we do for ourselves in order to start healing.

9. Pause, Count Ten

When we are emotionally overwhelmed and have run out of patience, it is important to pause before answering. As a good strategy to not escalate tensions, after the emotional circuits are triggered, take some time if it is safe to interrupt our own response. The idea is to slowly count to ten while waiting for the initial emotional reaction to pass. The goal is to exercise some self-control by calming down. While counting to ten, breathe deeply and maintain situational awareness by focusing on what is happening right now in this place and at this time.

All of this is easy to say, but there can be real challenges in doing it. It works because the foundation part circuits are very fast (about five milliseconds) compared to the upper part processing time (a second or two). Thus, the pause gives the upper part processing time needed, which may be important to override the foundation part, if it is not too late.

This idea of pressing the pause button is not new. Thomas Jefferson (1743–1826) was many things, including the principal author of the Declaration of Independence (1776) and the third president of the United States (1801–1809)[103]. On February 21, 1825, he wrote a letter to his nephew Thomas Jefferson Smith that became known as his Canons of Conduct. Canon number ten said,

"10. When angry, count ten, before you speak,
if very angry, an hundred."

This is the good way to slow down when experiencing a strong emotional reaction, such as anger. This method seems as important today as when President Jefferson wrote it. Using the pause to slow down is consistent with how the brain processes information. A good start is to go slow by counting to ten; then if needed, to one hundred. This natural way of building in a pause gives our Go Slow Thinking the opportunity to do its job.

10. We Get Good at What We Practice

Our current actions shape who we are tomorrow. If we practice hate, we get good at hating. If we practice cooperation, we get good at cooperating. This is because we get good at what we practice. A blend of the implicit and explicit functions is used to support the mix of the frequently repeated processes required. This applies to muscle movements needed for skills like typing, driving a car, and playing a musical instrument.[104] It also applies to verbal scripts, supporting our habits of mind. In turn, what we practice shapes our beliefs about life, preferences on foods to eat, or clothes to wear, as well as opinions about various social groups in our community and so much more. These beliefs, preferences, and opinions help influence attention's filtering action. Habits of mind easily recall certain scripts, which get repeated over and over again without thought. All of this impacts our relationships at home, work, and other settings.

From suitable feedback, we can make the choice to improve some feature in our life. With Go Slow Thinking, we can now choose our next move in this moment. Procedural memory, sometimes referred to as muscle memory, is needed to learn skills involving automatic repeated physical movements. Verbal skills get developed in the same way. We start with the intent to learn by engaging in the deliberate work to establish the new skill. The need here is to focus in the moment and regularly practice to develop any skill correctly. Procedural memory is created by practicing the necessary steps in the right order over many repetitions. These repeated actions create, strengthen, and help to condition the neural circuits needed to support this memory. The neural circuits are shaped by these repeated exposures leading to an accumulation of lots of small changes and tiny improvements. Over time this aggregation of small changes has an impact.

Regular practice then continues to improve and maintain this skill. Then procedural memory allows us to do the appropriate performance when it matters, accomplishing the results we were seeking. The results are motor skills, demonstrated as a change in behavior performed without any conscious recall like tying shoes, riding a bike, and brushing our teeth. This also includes our capacity to use specific verbal narratives in the moment to support our point of view. Habits of mind are developed and maintained in this way.

The results of procedural memory are the layers of experience gained by practice. The learned muscle memory accumulates, becoming an instinctual response when cued. The same pattern of routinely making this effort is used to learn verbal scripts available as needed when cued, becoming part of our automatic thoughts and words said to others. This is because, in the end, we get good at what we practice.

PROCEDURAL MEMORY

→ Used to learn a skill correctly.

→ Practice the steps in the right order over many repetitions.

→ Accumulation of lots of small changes & tiny improvements over time has an impact.

11. Improve Our Emotional Self-Awareness

There are many good sources available to help us apply the science of human emotions in our daily life. For example, emotional intelligence develops the capacity to recognize, manage, and improve our own self-awareness, as well as self-control. Emotional intelligence also includes empathy and social skills needed in learning to better handle relationships with other people.[105] A related approach is considering our emotional style in the areas of resilience, outlook, social intuition, self-awareness, attention, and sensitivity to context.[106]

The hard part is figuring out where to start. So, if we are unsure, here is a suggestion. Emotional self-awareness works to improve noticing the messages our body sends that impact feelings and thoughts. We can learn how to improve our own emotional self-awareness in three steps: (1) identify negative and positive emotions; (2) learn to assign a label to the emotional experience, as well as (3) learn to assign a level of intensity: red, yellow, or green.

The basic method to develop this emotional self-awareness involves using Go Slow Thinking to monitor the biological changes inside our body. At the core is the Jefferson method to press the pause button, take a deep breath, and count to ten. When it is safe, take the time to reflect upon the emotional state just experienced. The reflection work may seem like we are being a spectator of our own emotional selves.[107] During this pause, we can consider the various biological elements involved in a gut reaction like heart rate, breathing, and stomach. We can then assign just the right word(s) to the experience, and consider the intensity. Afterward, we can finish up by choosing our next move in this moment.

Identify Negative and Positive Emotions

Research has demonstrated how negative emotions are independent from positive emotions.[108] In addition, some research suggests having the goal in our daily lives to experience a 3:1 ratio of positive to negative emotions.[109] Positive emotions are experienced more mildly and frequently than negative emotions. The issue is being aware of positive emotions. This ratio of experiencing three positive emotions to one negative emotion is similar to how most healthy diets require a nice balance by routinely eating a variety of foods. So, a good starting point to improve our self-awareness is to notice positive and negative emotions arising within our body.

In business, a general quality improvement idea is if we can measure it, we can manage it. In following this idea, we can start working on improving our self-awareness by counting an emotional experience as positive or negative. The measurement element requires Go Slow Thinking to deliberately notice the experience of an emotional state and then consciously work to describe the body senses being experienced right now. From there we can assign if this was a negative and\or positive emotional experience.

Negative emotions narrow and focus upon danger or other threatening events demanding immediate attention in this moment in order to support survival-oriented actions. We need to keep our guard up to protect against dangers giving us a sense of unease due to a possible threat or something bad to avoid. Threat triggers survival mode to ensure safety needs are attended to first. The behaviors and thoughts triggered by negative emotions as the natural default setting are very energy-consuming.

Positive emotions broaden and build resources by opening our mind for micro-moments of expanded awareness to possibilities. Positive emotions give us a sense of ease to approach something good as an opportunity. This feeling of safety helps us to be comfortable in making the approach. Over time, positive emotions have cumulative effects by building up resources needed to support our survival. They are energy-nourishing and support prosocial behaviors. Positive emotions can easily be taken for granted, given their mild nature and the negative default. It takes the deliberate effort of Go Slow Thinking to develop the habits of mind to actively notice positive emotions. For most of us, we will notice more positive emotions in our lives than one may initially expect. The attention paid to the positive emotional moment can now be formed and stored in long-term memory, with cumulative effects.

Experience tells us how we can have mixed positive and negative emotions at the same time. Ambiguous events, memories, or imaginations trigger more than one emotional state, giving us a composite of biological signals.[110] This blend of signals leave us sensing mixed emotions. We are glad to have the job (need the money) but do not like some aspects of the work involved. Love living in this neighborhood but do not like the traffic in the area. As a player, we had a good performance (we did our best and were successful) but the team lost. We have all experienced the blend of happy and sad news at one time or another. We say people of good will can agree to disagree on something specific. This is all part of the effort in noticing our body's messaging when experiencing negative and positive emotions.

Improving emotional self-awareness involves picking up on our bodily sensations from head to toe, sometimes referred to as a gut feeling or a visceral reaction. We may be noticing biological changes in heart rate, breathing, or gastrointestinal issues. There may be behavioral changes as in energy levels, sleeping patterns, or muscle movements needed to support our actions. Notice changes in our thinking like concentrating, decision-making, or other routine mental tasks.[111] There is noticing the effects of Neuroception which involves the vagus nerve responding giving us a sense to avoid or approach an adversary. On the other hand, if we sense the place is safe and it is okay for social engagement, there is a damper on the vagus nerve, serving as a brake, inhibiting the sympathetic nervous system's influences setting up moments needed to socially interact with our near and dear friend. Quick Thinking's automatic, fast reaction with ease uses these signals routinely to carry out its role. We need to use Go Slow Thinking to deliberately and consciously recognize these changes in order to improve our emotional self-awareness.

We gain experience while moving forward in our efforts to identify positive and negative emotional moments. The novel first-time exposure may have been negative or positive, but we have trouble with the designation in the moment with no life experience to draw upon. Life tends to flow in patterns, repeated over and over again. As experience is gained and associations build up with these patterns, we now have knowledge available to better give meaning from the events, classifying and assigning the emotional understanding in our own way.

Learn to Assign a Label to Emotional Experiences

Building off these initial efforts, we can advance on to mentally labeling the specific emotional state. The idea is to learn to name the emotional state with the small modest goal of trying to improve our self-awareness. Because each brain is unique, we each experience negative threats and positive opportunities differently. The idea is to use our knowledge of emotional concepts to classify or categorize the current experience.

As we experience the emotional state, the initial focus is on consciously and deliberately describing the body senses felt right now as positive or negative. From there, we can use this additional label in our mind and when sharing our feelings about what happened with another person.

We can start with the traditional emotional language. With negative emotions, we can use words like fear, anger, sadness, or disgust. With positive emotions, the word choices include gratitude, elation, hope, amusement, wonder, inspiration, love, and joy. However, there are many more language options, such as:

Anger: Irritated, resentful, upset, mad, furious, raging, wrath, out of control.

Sad: Down, blue, pensive, mopey, grief, dejected, depressed, gloomy, heartbroken.

Fear: Tense, nervous, apprehensive, anxious, jittery, frightened, scared, terror.

Happy: Pleased, optimistic, satisfied, joy, glad, contented, serenity, fulfilled, ecstasy.

Gratitude: Thankful, appreciate the kindness, acknowledge of bless-
ings, obligation.

Hope: Optimism, good expectation, anticipation, better tomorrow, seeds for growth, fulfillment to be achieved.

Red, Yellow, and Green Zones

The next step in developing our emotional self-awareness involves considering the intensity of the experience. We can take the
concepts of a traffic light (Stop, Caution, Go) and a thermometer
(measure changes of temperature) to create a tool of measurement.
Red Zone—Danger, Yellow Zone—Caution, or Green Zone—Go
Ahead are useful to gage the intensity of the positive and negative
emotion. We use Go Slow Thinking to deliberately ask ourselves
questions about the emotional experience. After thinking about the
experience, assign the level of intensity.

Red Zone—Danger!
Threat triggers survival mode with ride, hide, engage. The Red Zone
warning applies to a variety of concerns. When safe, pause to consider these options.

Negative emotions

- Over the centuries people have had sudden, shattering, uncon-
trollable, overwhelming, life-threatening experiences. Fueled by
fear, the soldier's guard is up while sensing of danger from the
constant sounds, sights, and smells of war. Then the combat-
ant came home, describing the effects, using terms like soldiers
heart, shell shock, battle fatigue, and combat exhaustion.[112]
These expressions are now covered by the concept of trauma.

- We can feel distressed and overwhelmed from the ongoing fear,
sadness, and anger due to the forces pulling and pushing on us,
especially when the threat remains constant. We are scared all
the time. Is living under this constant threat the new normal?

- Medical illness can take away our abilities to care for ourselves and family.

- In business and romance, when the pattern shows actions do not match the words, a red flag in our mind goes up. Now we need to deal with a person who lies, cheats, and cannot be trusted. They do not honor their agreements. They lack the courage to acknowledge when wrong. They use disinformation to deceive. They dish out unwarranted feedback and ignore any suitable feedback.

Positive emotion—Compulsive use despite consequences.

- We, or someone we care about is caught up in the compulsive use of rewarding stimuli despite consequences (alcohol, drugs, gaming and more).

Yellow Zone—Caution

Neuroception is monitoring for unknown risks; provide a warning on any imminent threat, then pause to determine avoid or approach. The yellow zone caution covers the concept of neuroception, the natural negative default monitoring for threats and opportunities. Keeping our guard up is smart as we deal with first-time novel events. With experience, we learn what to avoid or approach.

- Nobody is perfect. We made a reasonable effort, but errors were made, and our actions came up short. Included misinformation was accidently passed on. Use suitable feedback to make appropriate adjustments and keep going.

- Feel worried, frustrated, nervous and not sure what to do. In our mind, the be-careful sign is flashing. Use caution; proceed when ready.

Green Zone—Go Ahead

The Green Zone is a place we work to get into with a nice steady flow. Here rest and digest return us to homeostasis. We want to maintain the flow of this healthy balance.

Negative emotions

- With eustress, we can meet the challenges. We have the physical resources, the mental focus, and the determination required to prepare and engage in the challenges.

- We dedicate ourselves to take the time and focus on the goals in order to succeed.

- Real external deadlines to take action (finish time, drop-dead deadlines; down to the wire) are effectively used to prepare and engage behavior needed to play the game, do the performance, make an application for school, work, and finances.

- This is the grit, backbone, fortitude, and determination needed to keep going.

Positive Emotions

- People do what they say and say what they do. Agreements are honored. Trust is earned.

- Enjoying the rewarding stimuli while knowing how to use responsibly within limits.

	Negative (–) Emotions	**Positive (+) Emotions**
Red Zone— Danger!	Ride, Hide, Engage	Compulsive use despite consequences.
Yellow Zone— Caution	Neuroception, guard is up. Monitoring for unknown risks; Provide a warning on any imminent threat. Pause to determine avoid or approach.	
Green Zone— Go ahead	Eustress, grit, backbone, fortitude, determined to keep going.	Enjoy, Use responsibly, Know the limits

Sometimes we know immediately the nature of the emotional experience. Other times we have mixed emotions about what just happened and need some time to think. There may be more than one way to look at events, with some unclear features making for ambiguous feelings. If we are not sure, when it is safe, pause, take a deep breath, and ask ourself some questions about what happened. This may be a time and place to pace ourselves, taking one step at a time, while trying to stay focused in the present moment. Right now, ask ourselves:

- How are we doing? What is going on? Where are we? What are we trying to accomplish? How did this happen? Why?

- Were the bodily senses being experienced right now positive (+) or negative (–) emotions?

- What grabbed our attention triggering the emotion? The context matters so it is important to pay attention to what the triggering event(s) were for the emotion. The situation includes external content hitting our senses (light, sound, smell, taste, touch) and internal self-talk.

- What actions took place such as body movement, facial expressions, and words used? What opinions, judgments or decisions were made? What were the consequences? How was the experience, Good (to be repeated) or Bad (to be avoided next time)?

- The words to use in describing the experience are (then say the words).

- The intensity of the experience was red, yellow, green.

- Focus and think about our next steps carefully. Proceed when ready.

Our past influences the current emotional experience. Quick Thinking in the moment is influenced by what passes through the attention filters with long-term memory helping to inform the moment. Attention filters prioritize the incoming content to narrow down to the one point of focus. Long-term memory is reconstructed at the time of recall using stored bits and pieces selected to support the efforts in the moment. Meanwhile, the other bits and pieces from the past are ignored, not needed in this moment. So, our unconscious interpretation of the emotional event is biased by our own attention filters and long-term memory.

How were we taught to react? Our parents, friends, co-workers, the people living in the community, and the media teach us concepts to classify or categorize the emotional event. People around us say, "Hey, didn't their actions make you angry?" or "Oh wow, that must have really scared you, huh." or "it is a sad day when that happens" or "yes, what a happy ending" and so on. In time these pattered expressions are repeated again and again helping us build associations reinforced by the people around. These labels help give meaning to us as well as communicate the experience to others.

As we work to understand the intensity of the emotional experience, remember: not all positive emotions are good; not all negative emotions are bad; we work to do our best under these circumstances; we can pick our battles, and look out for the booby traps set to create gotcha moments.

Each step within learning emotional self-awareness offers opportunities to make changes and personal adjustments. We have been working on improving our emotional self-awareness by identifying negative and positive emotions, assigning words to these emotions, and rating their intensity. These fundamentals building emotional self-awareness are useful to learn but take time and effort. By taking one step at a time, we will see improvement over time with the accumulation of many small efforts repeated over and over again.

With those choices, we can use this insight to tame the emotion by toning down and softening the impact, then over time bring it under better control. Developing the habits of mind to recognize the context of the emotional state takes time as influenced by intensity of the event and the words chosen to describe what happened. As in so many things, we get good at what we practice.

12. Concept of Empathy

Empathy is built into our biological core, starting as the simple idea of unconsciously sensing other people's emotions.[113] We vicariously experience another person's emotions without having their thoughts and feelings explicitly communicated.[114] Empathy certainly has a reputation for kindness, understanding, and positive support toward another, such as we may see between a parent and a young child. But things are not that simple. Empathy can be very painful when the emotions in other people involve distress due to fear, sadness, and anger.

But empathy is much more complex than just reading emotions. Empathy starts with a simple operation at the core, building out into something complex and nuanced, highly dependent upon the specific context. The part of empathy involving one person unconsciously picking up on another person's emotions can be used for positive support or to exploit a victim. The con man can use empathy to manipulate us into agreements, such as purchasing products we do not need or to cast our vote in elections against our own best interests.

Empathy is part of emotional intelligence, involving the recognition of emotions in other people, an important tool in social competence and managing relationships.[115] It is the automatic and unconscious collection of emotional information about other people.

Improving our understanding of empathy will help with two of the six dimensions of emotional style: social intuition and sensitivity to context.[116] Social intuition involves how well we read other people's affect, as in tone of voice, facial expressions, body language, and other outward indicators of emotions. Context involves how well we learn the social rules within various groups and cultures.

There are three layers to describe the various parts of empathy: emotional empathy; cognitive empathy; and compassionate empathy.[117]

Emotional empathy is our biological core, working as the feel-it layer. Here our core biological functioning starts the unconscious and automatic process of matching up our individual emotional state(s) with another person's affect. Emotional empathy is where we are sharing another person's feeling. This is our foundation part's nonverbal, unconscious biological core picking up on and matching this other person's emotional state, referred to as emotional contagion.[118] As a result, we can "catch" each other's emotions.[119]

Cognitive empathy is the think-it layer, involving awareness. Theory of mind is the capacity to take perspective by discerning and understanding another person's feelings and thinking. Here we are thinking about the other person's point of view, including their emotions. We show the capacity to understand how the emotional, cognitive, and other mental processes of another person can be different from ours. This information on mental processes of other people can be used to help anticipate another person's actions and explain their behavior. We notice the other's emotional state and make predictions on their intentions.

Empathy is needed for successful group living. This capacity for cooperation and social behavior gives the parents motivation to attend to the needs of their children. It gives partners the mental capacity to cooperate with each other. This helps members to signal about pending dangers, provide assistance to others as needed, organize activity, be sensitive to each individual's emotional states, mimic actions, synchronize behaviors, and more. We use this biological capacity for empathy to engage in the cooperation, and social behavior.[120]

Compassionate empathy involves being moved by our observations, triggering the need to take action. Empathetic concern, or compassion, motivates one person to have a desire to help another person in need. The compassion toward members of our group motivates us to help. Empathetic concern can be very painful if watching other people suffer. It motivates parents to help out their children, families, and friends as needed. People who work in helping professions like health and human services, law enforcement, firefighting, emergency medical and related efforts are routinely exposed to other people's suffering. The ongoing exposure can lead to cumulative stress and burnout or compassion fatigue if not properly managed.

Three Layers of Empathy

Emotional Empathy	Biological Core	The feel-it layer
Cognitive Empathy	Awareness	The think-it layer
Compassionate Empathy	Action	Be moved by it layer

Emotional Support

Empathy is needed to sustain emotional attachments between individuals as two unique brains form one unique relationship. Keeping a successful relationship requires a commitment of time and effort by both people involved. Empathy gives romantic partners, teammates, and co-workers the ability to cooperate with each other as well as pick-up signals when one needs assistance.

Emotional Support is a micro level person to person effort. The support can start as simple practical assistance like offering some water to drink. It can then involve listening to the other person's story in a safe, non-judgmental way. It may also include providing a shoulder to cry upon, having each other's backs while facing the outside world, as well as offering suitable feedback in the privacy and safety of our shared place.

Being there for each other takes it all a step further. The two people involved have made a special sustainable connection. We take the time to really listen to each other. We work on having clear, consistent and honest communication on a regular basis. No matter what life may bring our way, our goal is to provide, care, love, and support to each other. We are unconditionally positive regarding each other with outside people. We learn how to avoid the previous wounds and triggers from prior relationships.

The relationship needs to be a priority for both parties in order to strike the right balance and manage the inevitable complications. There is no one size fits all. It comes down to how the ordinary everyday interactions work between two people. Do we feel listened to in these moments? Is there emotional transparency? When together, do we feel safe?

Empathy helps set up another form of emotional backing called peer support. It can help people face the unknown by hearing how others experienced a similar event and receive some basic education. This is where one group member listens to another member's stories and provides a shoulder to cry upon. Peer support is based on the common theme of the shared lived experiences.

The compassionate empathy naturally flows from a special connection to another person who has gone through a similar challenging experience and survived. The peer offering the emotional support "gets it," knowing the feelings involved from the lived experience.[121] Peer support happens when one person with past common experience provides practical assistance to another individual who is currently overwhelmed, confused, upset, and struggling. This is part of a distinctly human feature of our natural drive as social beings, as we want to swap stories and exchange thoughts with other people.[122] The peer offers the emotional supports by reaching out, making themselves available to just listen, and, if requested, providing practical guidance.

The peer may be a veteran soldier helping one who just came home from a war zone, bereaved parents who lost a child to gun violence, one who experienced a serious medical condition and has recovered, a co-worker in a challenging assignment, or many other combinations.[123] Support settings may range from informal situations like sharing a meal after work, up to structured programs provided by trained peers with specialized knowledge and skills to systematically help to others cope with the stress of a new normal.

Peer support can be provided at a variety of times, starting with the immediate aftermath of the challenging events. Additionally, it may be offered during the inevitable times after the world moves on and certain disillusionment sets in as one is feeling lonely and overwhelmed by the impact of what occurred. The anniversary of the event as well as other significant dates like birthdays and holidays may be especially painful times. The key part of compassionate empathy via peer support is more meaningful when it comes from someone who has had a similar experience.

The peer needs to have experienced a similar struggle. It will not be the same experience because each human brain is unique. The similar relevant experience means the peer can say,

"Yea, I have been there, done that, and felt that way."

"I have been scared, angry, upset, overwhelmed and sad just like you are now."

"I faced the similar dangers, survived the trauma, and overcame similar types of challenges."

As a result, the peer can relate to this other person who has recently survived a similar situation. With the lived experience in common, the peer can serve as a role model on how to cope and recover for the one now struggling from the recent experience.

Peer support models have been used for many years in a wide variety of settings. The Alcoholics Anonymous program has offered peer support to people with alcohol addiction problems for decades. More recently peer support has emerged as an important organized service in community mental health. Peer support is an important feature of the Critical Incident Stress Management program, which provides assistance to first responders like law enforcement officers, fire fighters, emergency medical personnel, and others. Peer support models have been adapted to assist many other important concerns. Empathy drives us toward wanting to offer compassionate care to others who are struggling. Talking to a peer with a similar lived experience will often help a person better cope immediately and in the long term, reducing negative emotional consequences.

Another form of support is an after-action talk session. Here the team leader in charge of the group organizes an informal support and feedback discussion between the members after their work is done. The goal here is to take the time to review, comment, and give feedback on what went well with this assignment, the challenges that occurred, and ask if one thing could be changed, what would be it? The talk session process is designed to promote cooperation by having the workers review the assignment just completed. In this respect, the discussion can help us keep everything in perspective. Beyond the emotional release as part of a cathartic experience, after-action talks between peers, co-workers, and colleagues can build relationships, helping to promote teamwork. The talk session can help reduce the emotional impact from an especially challenging, overwhelming event connected to the work assignment.

Vicarious trauma is a concern for people providing peer support to individuals who have survived dangerous life-threatening events. Listening to these stories has its own cumulative impacts, leading to compassion fatigue. It is important to learn the self-care measures of stress management to prevent burnout from offering these positive empathic supports.

XII. ETHICAL PRINCIPLES

Emotions give meaning in the moment, which includes our sense of right and wrong, leading directly into the concept of *ethics*. The word ethics refers to one's personal code of conduct as seen in actions, behaviors, plus words demonstrating our principles, standards, and manners.

Ethical actions start inside the mind with how our emotions influence the processing of information, providing us a sense of right and wrong. At the individual level, the reality of culture helps to define what is right and wrong consistent with group concepts of threats or opportunities. At the collective level, the dynamic leader teaches what is right and wrong for members inside that group as well as how to deal with outsider relationships. A chain of bias transmission can be one tool for supporting this leader's lessons.

We need to make the choice to learn and practice ethics as part of our problem solving and dispute resolution efforts. The verbal scripts associated with our sense of right and wrong are useful for making our way through this complex and contentious society. For people working to solve problems, improve relationships, as well as resolve disputes, knowledge and practice of ethics are essential tools to accomplish those tasks.

Part of providing quality services includes ethical principles like:

- Do no harm; try to make a positive difference;

- Know our limits; confine practice to areas of expertise. If we do not have it, learn it. Acknowledge our errors;

- Work to ensure the safety of the people involved, starting with the person served;

- Treat people with dignity and respect. Start with setting a good example by showing respect first, be in the moment without judgment, and treat other people the way we like to be treated.[124]

- Do a little CYA (Check Your Assumptions) when working with people different from ourselves. It starts with acknowledging each person has a unique brain. There are many differences based on values, opinions, and attitudes; as well as economics, age, gender, race, color, national origin, religion, and health status.

- In the spirit of self-determination, work toward reaching a voluntary uncoerced agreement to settle disputes by respecting individual capacity for decision-making. This may range from the ones able to make their own choices to those who are vulnerable and need proper assistance due to health impairments.

- Rights and responsibilities are two-sides of the same coin. For each individual right, there is a corresponding responsibility to the collective common good as a member of the group;[125]

- Good intent is nice. However, our words and actions will be the basis used by others to judge what happened.

- Remember neuroception is the continuous, unconscious monitoring for dangers and threats. A trauma-informed approach to providing services involves awareness of the lasting adverse effects from life-threatening events.[126]

- Within any dispute, the people involved have emotional stakes, including possible financial interests influencing their judgments, choices, and actions taken. *Conflicts of interest* are identified when the assigned neutral facilitator has an emotional or financial stake in the dispute. The neutral facilitator needs to acknowledge any connections, which may compromise their role to those involved in the dispute and step away if anyone raises a concern.

Ethics are part of good social skills used to facilitate interaction and communication with others, both in verbal and nonverbal ways. Personal responsibility flows into our obligations to sustain relationships at home, work, and other social settings. Ethical principles will be followed if routinely reviewed and the leaders expect adherence. Such expectation and learning will move these ethical principles into our habits of mind as part of the routine ways of thinking shaped by one's experiences.[127]

XIII. SCIENCE INTO ACTION

Today we live upon the mountains of accomplishments and follow the pathways forged by our ancestors. The leaders of our ancestors' clans supported the discoveries of their day. After all, those discoveries made group survival easier. Pathways shaped by trial and error slowly formed into more orderly efforts. As the work of discovery became better organized, the scientific method started to emerge.

The scientific method was developed over the centuries to reduce the subjective features and increase the objective elements to reveal the realities of how the world works. However, the results of these efforts have not always been what others wanted to hear. For example, just consider what Nicolaus Copernicus (1473–1543) faced in reporting his findings of how Earth orbits the Sun while rotating daily.[128] Thomas Willis (1621–1675) pioneered research on the human brain, starting the era of modern neuroanatomy and, in the process, helped to improve upon the scientific method.[129] Reporting the facts on the reality of physical elements to a skeptical audience is no easier today.

Scientific methods have been used to research emotions. What has been learned over the years about our human emotions is only useful if we take the time to apply this knowledge by putting the science into action. The science is the biology of emotions. The action is understanding how emotions influence our choices, attention, thinking, and behaviors. With this understanding, all people can improve their ability to solve problems, resolve conflicts, and have better relationships. After all, our emotional brain is where all conflicts begin and end.

Head → Objective – reduce the distortion from emotions. The word *objective* suggests dealing with conditions perceived as being without distortion by emotions, feelings, bias, or prejudices. Facts can then be established, using deliberate methods to show something actually physically exists in reality. These efforts work to reduce subjective influences to a minimum.

Heart → Subjective – emotions influencing preferences. *Subjective* suggests positive and negative emotions impact how we evaluate options, leading to favoring particular ingredients, features, or other qualities used to make judgments. These preferences are not necessarily based on reality or having substance.

Physical Elements are what happens in fact, such as the composition, interactions, processes, and known properties of what is around us.[130] Physical elements of reality do not change. Human understanding of these physical elements does change by using the scientific method. For example, science established the reality of how the Earth orbits the Sun while rotating.

Social Features are a result of the human mind interpreting physical elements. It includes people agreeing on the words used to describe and give meaning to the physical facts. From these interpretations, we learn concepts. For example, social features created the designation of time it takes the Earth to orbit the Sun as one year and the Earth completing a rotation as one day.

XIV. SOME FINAL COMMENTS

By putting this science into action, we can start the work on improving our knowledge involving the emotional side of conflict. In this case, the action represents understanding the universal biology of human emotion's influence on our attention, thinking, actions, and decisions.

Conflicts and disputes driven by emotions based on our universal, yet unique, brains are just part of living. With this knowledge of emotions, we can choose how to work on resolving these conflicts. There are more options than engaging in never-ending disputes, using unrelenting harsh rhetoric. Disputes can range from a mild difference with another up to an armed violent struggle between rival groups, including overt acts of warfare between nations. But in the long term, the violent approach is very hard to sustain. Most of the time, finding a lasting, sustainable settlement requires the two conflicting sides to start talking. Many dispute resolution approaches are based on various communication strategies between the people involved with each one emotionally reacting, disagreeing, and then in calmer moments eventually agreeing to talk.

As part of coping in the moment, we have to work with what we have. We can start by developing the practice of acknowledging our own emotions, using this special language. Then we continue by expanding our knowledge of emotions, including the special words, concepts, and expressions. giving Go Slow Thinking increased options available to work with in the moment.

We get good at what we practice. With the effort invested in learning, we can work on finding the right road moving in the right direction. The right road is being familiar with the basic science of how emotions function. The right direction involves learning to use this science to better our lives in order to improve relationships and reduce conflicts.

ATTACHMENT A: HUMAN NEEDS

These are good ideas to keep in mind as we work to solve problems, improve relationships, and resolve disputes. They may also be useful in stress management as well as aspects of dispute resolution when considering interests, issues, and positions.

Maslow's Hierarchy of Needs[131]
When taking stock of our life, Maslow's Hierarchy of Needs provides a nice easily accessible tool to sort out in priority order the interests we want to satisfy. Here we test our list of the "vital few" from the "trivial many" activities into priorities based on: (1) Physiological needs; (2) Safety needs; (3) Acceptance needs; (4) Esteem needs; and (5) Self-actualization needs.

1. Physiological needs—In general, our human body's fundamental biological needs are the same universally around the world including air with oxygen, food, water, and shelter to sleep. If these human biological needs are not met, the person does not survive.

2. Safety needs—After the physiological needs have been more or less addressed, then security issues are next in line. This includes having predictable, stable lives, including meaningful daily routines (such as a job, school, or caretaking), reasonable control to make choices, financial security, safety addressed at work and home, law and order, access to health care, literacy, as well as prevention of injury and accidents.

3. Acceptance needs—after physiological and safety needs have been addressed, social needs including interpersonal relationships, including love and belonging can motivate one's actions. Cooperative efforts are built on trust, leading to friendship, giving and receiving affection, establishing households, group membership, and more.

4. Esteem needs—after physiological, safety, and acceptance are esteem needs. Here issues like dignity and respect become important as well as mastery. Self–esteem and reputation are important.

5. Self-actualization needs—After the person's deficiency needs have been addressed (physiological, safety, acceptance, and esteem), it makes possible for one to work on the growth goals at the self-actualization level. Here the positive accumulative influence of successfully addressing deficiency needs builds up a foundation, allowing a person to try to follow and achieve their interests to whatever degree is possible. This growth includes athletic, artistic, academic, economic, or other pursuits as the person has time, interest, and capacity to engage.

However, the deficit needs are always present so if the lower levels get disrupted due to how life can happen, such as job loss, divorce, natural disaster, accidents, and more, a person returns to lower levels of this hierarchy.

Individual survival requires a person's deficiency needs be addressed, starting with clean air to breathe, adequate food to eat, good water to drink, sufficient sleep, and avoiding predators. Here are several additional physiological and safety needs to consider.[132]

HOUSING—a stable, suitable, decent, affordable, safe place to live, with properly functioning plumbing, electricity, and temperature control. It is a place to call home. People with few resources experience housing cost burdens, as well as food and water insecurity.

CLOTHES—Suitable clothing for the environment where one lives and works.

TRANSPORTATION—suitable means to get around within the community where one lives. In metropolitan areas, public transportation options are available. In suburban and rural areas, automobiles are essential.

HEALTH CARE—access to suitable health care is important. Then as needed, making informed choices in the treatment for illness, followed by learning how to manage one's symptoms and work toward recovery. In the public health sense, effective approaches to prevent the spread of disease benefit all residents of an area. In addition, healthy workers at all levels are needed for a sustainable society.

INCOME\EMPLOYMENT—In today's developed world, no one survives as hunters and gatherers like our ancestors did. Also, very few can survive as self-sufficient farmers. Thus, one needs to have a source of money to purchase food, water, clothing, shelter, transportation, health care, food, and other items. Most of us need to earn the money through employment. The work needs to pay enough to purchase what is essential for survival.

DIGITAL LITERACY—this area starts with education, including the ability to read. Today this also needs to include understanding a basic use of computers. At minimum, dependent populations (people who are children, elderly, and/or disabled) need to have someone who can represent them in the digital world.

RELATIONSHIPS—as people, we are interdependent upon each other for friendship, support, and love. To develop and sustain these relationships, everyone needs to learn basic social skills (including emotional intelligence) and problem solving. These techniques help individuals to form and sustain relationships needed for coping in this complex world. Using these skills can help establish the single greatest hurdle between people—trust. Honoring our agreements and doing our part in the relationship is essential. Without these skills, to sustain the relationship between individuals is difficult.

ATTACHMENT B: TOOTH TO TAIL RATIO—WHOLE BODY

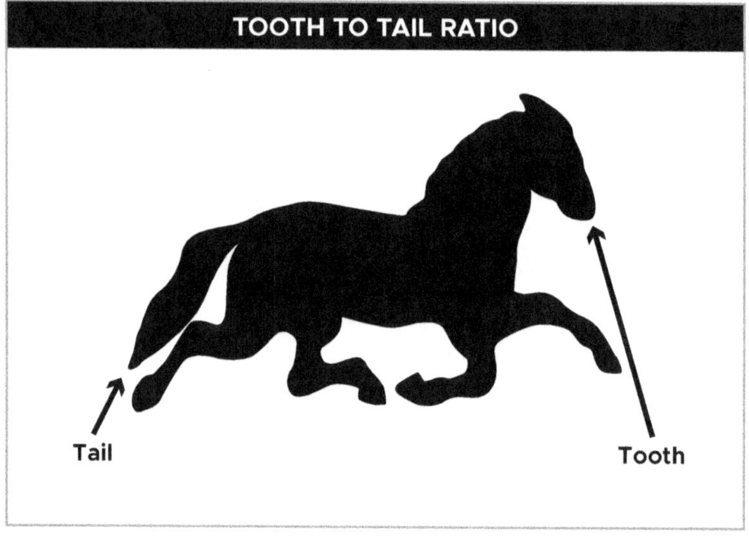

In a famous scene from *The Wonderful Wizard of Oz*, Dorothy, Scarecrow, and the Tin Woodman are clearly scared while walking slowly down the yellow brick road in a dark, creepy, thick woods. They feared a possible encounter with growling wild animals who put their teeth in the right place to capture food. When they do encounter the Lion, at first, he seems frightening. After Dorothy slaps him, he confesses to a need for courage.[133] This little scene reminded me when a speaker talked of the tooth-to-tail ratio. The speaker said it takes the whole animal's body, from the tail forward to put the teeth in a position to grab some food. If all of these body parts are not working together in the moment, the animal goes hungry.

There are many examples of a tooth-to-tail ratio in daily living, including film credits, space flight, the aircraft carrier, and team sports. Have you ever noticed film credits? It is incredible the number of people required to make one film, even with those using only a couple of actors. It took about 400,000 people working in the space program to land twelve men on the moon between 1969 and 1972.[134] An aircraft carrier is a key type of ship in our military naval force. One aircraft carrier can have over 5,000 personnel, supporting sixty planes.[135] This does not count the support ships accompanying the aircraft carrier or the efforts needed to train the pilots, build the planes, or design the ship. In team sports, there are key positions needed for success, such as quarterback, setter, goalie, pitcher, or point guard. But it takes a variety of skilled players to make the team successful, not to mention coaches and others handling off-field details.

<table>
<tr><th colspan="2">GUT REACTION (UNIVERSAL – ALL)</th></tr>
<tr><td>Foundation Part</td><td>MANY PARTS OF BODY INVOLVED WITH EMOTIONS</td></tr>
<tr><td>

- Amygdala
- Hippocampus
- Autonomic Nervous System:
  • Sympathetic Nervous System
  • Parasympathetic Nervous System
- Hypothalamus-Pituitary-Adrenal Axis

</td><td>

- Prefrontal Cortex
- Insular cortices
- Reward Pathway
- Periaqueductal Gray
- Cranial Nerves
- Microbiome
- Language

</td></tr>
</table>

So is the case with emotions. The emotional system is an important set of functions used by our whole body, including the brain, for survival. There are a number of biological structures involved in giving us emotions. Many descriptions start with the amygdala. Actually, there is an amygdala on the left and right side of the brain so the plural, amygdalae, is more appropriate. However, the common practice in discussing this part of the brain is to use the singular, amygdala.

The amygdala is heavily involved with emotions and memory. However, emotions are not a solo function of the amygdala. In addition to the amygdala, there are a number of other sections within the brain and body, which have important roles. The hippocampus supports new memory, storing explicit emotional experiences, and helps us to navigate around. The autonomic nervous system has the sympathetic nervous system (fight or flight as part of the stress response) and the parasympathetic nervous system (rest and digest). The prefrontal cortex helps to regulate emotions by quieting down the amygdala, decision-making, impulse controls (Go\Stop), sound judgment, planning, plus other executive functions. Insular cortices monitor the internal visceral organs. Specialized neural circuits help to support the capacity to read the muscle movements in another individual, such as the emotional expressions on other people's faces. The hypothalamus-pituitary-adrenal (HPA) axis is part of the stress response needed to react to threats. The reward pathway associates useful behaviors with pleasure, such as seeking food or finding a mate. The periaqueductal gray is found in our brain stem, helping with the freeze response to threats and to cope with pain. There are twelve pairs of cranial nerves connecting directly from our brain to bodily functional areas like smell, vision, taste, sound, facial expressions, heart rate, breathing, vocal cords and more. The microbiome involves the microorganisms living in and on our body, providing important support functions. Then there are the language areas in the brain, providing emotional words, self-talk, and more.

It all comes together in our emotional system. Emotion is part of every unconscious action and conscious thought. There are physical microscopic level events scaling up through our neural circuits to provide a subjective sense of pain (–) and/or pleasure (+) that is either present or absent. Emotional memory is there, unconsciously and seamlessly influencing our thinking. Some of our experiences include feelings when we sense the emotions playing out within our body.[136]

Our emotional system is built in layers with a number of functions and has an enormous impact on how we live. Learning to manage our emotions has always been an essential skill for success in life. Without developing this ability to manage, an immediate emotional response can happen in the blink of an eye with the resultant outburst we regret later.

Emotions are one of the more complex functions within our human body. All thoughts, emotions, physical movements, and more are managed by our human brain.[137] Within our body, the brain uses a disproportionate share of the available resources. The return on this investment is self-preservation, human connection, built up resources, and more.

ATTACHMENT C: BIASING–A FEW CONCEPTS TO CONSIDER

There are a variety of ways used to describe bias including:

Affinity Bias—As part of the us-versus-them dynamic, we find attractive features in other people who are similar to ourselves. We are not going to be automatically inclusive due to neuroception, the continuous unconscious monitoring for dangers and threats, creating a natural negative bias. The threat assessment done by our attention-filtering action notices skin color, gender, height, weight, and age very quickly. Is this person safe? The natural filtering action creates blind spots as we may perceive other people like ourselves to be safe while those who are different a threat. In the moment, we sense how another person who agrees with us seems to be especially good, which in turn unconsciously influences our choices. The skew becomes self-maintaining as fewer of "thems" seek membership to be inside our group.[138]

Attitude—A form of bias where emotional memory as well as the attention-filtering actions set up a predisposed appraisal toward certain people, events, issues, places, objects, and more. This habitual inclination quickly classifies the focus of attention with fast automatic thinking supporting our actions and narratives.

Availability Bias—What we are familiar with will come to mind faster. It is readily available for our use in this moment. Our brain automatically makes associations as it encounters related words, ideas, pictures, or experiences with a shared topic. In a discussion, we will say, "Oh that reminds me of ___." We all do this. Things can become familiar when our estimate of risk is distorted by the amount of coverage something gets via television, radio, social media, and related sources. Such coverage leans toward the novel items with deeply moving, emotionally painful topics. This same availability can be planted in our mind via the priming and anchoring functions. Our experience of the world is not an exact copy of reality. A pattern of frequently repeated messages can plant all kinds of ideas in our mind. This ease of the more familiar coming to mind can lead to good choices, little minor problems, on up to major errors with a great cost.[139]

Blind Allegiance Bias—Group loyalty, sustained by the beliefs in the claims and support of the mission-driven vision without question even in light of strong evidence to the contrary. The commitment is so strong the loyalty remains, even in light of incomplete, confusing details as well as overwhelming contradictory evidence.

No matter the evidence presented, the blind allegiance remains: Members simply ignore criticisms from the outside, fail to hold leaders accountable, and feel justified to spread disinformation in support of their cause.

Blind Spot Bias—is where we think of ourselves as less biased than another person.[140] This biasing function is a universal function inside all of our brains.

Blue Lie Bias—In a hyper-polarization, dualistic thinking, tribal us-versus-them atmosphere, lying can thrive as part of the group identification. A lie is an untrue statement or omission of content intended to deceive the listener. A little white lie is commonly understood to be a harmless falsehood to spare another person's feelings (such as "Yes, dinner was good tonight"). A blue lie is different. The blue lie is intended to pull our group together while attacking the other rival group(s). The emotionally charged atmosphere is full of our resentment and anger toward them. The followers overlook lies made by our group members. Meanwhile, these same followers are outraged by the errors and falsehoods the opponents said. In managing the adversarial relationship, the leaders carefully use the lies as weapons in the zero-sum game of a win\lose competition.[141]

Confirmation Bias—is the inclination to attend to information consistent with our own preconceived notions. Confirmation bias means as we pay attention, the filtering action narrows our focus to information supporting our own ideas and ignore or discount evidence to the contrary.[142] People will stubbornly reject any information inconsistent with their point of view. They will be highly critical of facts not supporting their position. They will stubbornly maintain how right they are and how wrong the other person is, even in the face of overwhelming evidence. This can move into willful blindness.

Consistency Bias—our ideas about ourselves lead us to revise our memory of the past to be very much like or very different from the present.[143]

Contempt Bias—Feeling superior in rank above another person, leading to disdain and a lack of respect toward another person or the actions of the person. This is bias against something or someone.[144]

Egocentric Bias—the powerful functioning of our brain to think highly of ourselves, with unrealistic perceptions and memories of our abilities.

Gotcha Moment Bias—Here we assume the worst and work to catch someone doing or saying something consistent with that theme. We are focused on looking for something wrong, disturbing, or embarrassing. When found, we proclaim we have got you—gotcha! Once caught, the action can be quickly shared with this individual and others. The shared thought sounds like, "I knew someone like you would do something like that!" The action can easily be blown out of proportion.

Herd Mentality Bias—Believing things because other people appear to believe the same thing. Inside the safety and support of the group, we let our guard down. It is not a time to question. This is basically where we go along to get along by saying, "Yeah, me too!"

As social beings, group membership is an important priority emotionally. We are members of many types of demographic groups, starting with simple things like gender, race, ethnicity, geographic location, education, and economic status. Then we have social groups, giving us the desire for conformity in various settings. Sometimes it becomes a matter of picking our battles, thus going along for now.

This is based on the idea that content repeated often and long enough is hard to separate from what is true.[146] Our Quick Thinking finds it easier to go along. This could also be called the bandwagon or groupthink bias. The leader says jump, and the follower just asks, how high? The follower is happy to go along instead of saying hell no!

Hindsight Bias—This is the "I-knew-it-all-along" type of thing. We use our current updated memories to help reconstruct past events. The emotional state's implicit bias helps to ignore some inconsistent emotional content. Forgetting and\or distorting other details add to the mix. Then it can feel like we were in a position to correctly predict what ended up happening.[147]

Mob Mentality—Herd Mentality can easily turn into a mob mentality fueled by emotional contagion as part of the us-versus-them dynamic. At the sporting event, musical concert, or political rally, we feel emotionally connected to the crowd. Acting on gratuitous negativity, the emotionally driven crowd of people starts to destroy someone or something. In the excitement of the moment, individual members of the group may take actions they normally would not do when alone. A lynch mob sets out to kill someone by hanging without a trial. An online lynch mob gets caught up in the moment, anonymously bashing someone or something. Racism 101 is to turn our opponent, the subject of "them," into a nonhuman\subhuman object. This is an important tool in us-versus-them.

Matilda Effect—the bias against acknowledging contributions of female scientists by ignoring, dismissing, or otherwise sidelining the work. Frequently the credit is attributed to a male associate.[148]

Racism—bias based on the idea there are differences in people based on race and to place the designated group into a subordinate position.[149]

Recency Bias—assuming the future will be like today and the recent past.[150]

Sexism—bias based on the person's sex, leading to stereotype expectations on social roles and to place the designated group into a subordinate position. Misogyny is sex discrimination against women.[151]

Sincerely Held Beliefs—are habits of mind developed over time by frequently repeated processes which form the implicit biases held. These sincerely held beliefs will sway our attention, memory, actions, and thinking. Some of these are useful to meeting the demands of living regarding business strategies, political movements, and religious traditions to be followed or ignored. Some may be criticized as prejudice. Supporting a point of view with sincerely held beliefs suggests one is closed to changing a fixed, static way of thinking, making such topics nonnegotiable.

Smartphone Bias—The information received via the electronic device seems to have higher value than other sources. This includes the need to constantly monitor the smartphone, and be alert, watchful, and maybe vigilant to content coming in on the device. It includes the claim one can successfully multitask, even while driving, at the dinner table, or in bed. Those without this bias know when it is a good time to use the smartphone device, and when not to, such as driving, meal time shared with others, and other face-to-face times. It includes bedtime, the time and place for sleep and recovery from the day.

Status Quo Bias—There is a willingness to ignore a problem or conflict with the hope it may go away. This includes the denial of problems leading to inaction based on liking things to stay basically the same. There can be a refusal to face difficult problems with the silent support for the current ways. This can be seen as willful blindness.

Stereotypical Bias—This is basically just lumping all who appear to be in a group into one generic category, usually negative. It is how our generic associative memories help to shape perception, leading to a fast automatic quick impression, which may be imperfect and can lead to systematic errors.[152]

Stigma—This bias is where one group looks upon another, the target population, in a negative way due to some reason like skin color, religious beliefs, a medical disorder, or other noticeable features. Stigma is a powerful force exercised in many subtle and overt forms. Discrimination is when this one group acts on the stigma by treating the target population differently. The negative treatment usually makes living life much harder for the target population.

Truth Bias—We tend to believe what other people say or write, allowing our relationships with family, friends, business associates, and others in the community to be easily engaged. We want to believe what other people say as the default setting in face-to-face and computer-based social interactions.[153] This is an important assumption because it would be difficult and time consuming to fact check everything we read or hear. We want to believe our partners, so in established relationships, small inconsistencies are overlooked or excused unless there is a reason not to.

Since some people do try to deceive, when encountering a stranger or preparing to spend money, it is understandable to have our guard up with some skepticism, uncertainty, and doubt in order to avoid being cheated.

Tunnel Vision Bias—this is an old phase, suggesting one is so narrow-minded due to tightly focused eyesight that much of the peripheral content is lost.[154] The old phase *tunnel vision* covers the filtering action of seeing, but the attention-biasing effect also includes the other senses of hearing, taste, smell, and touch. This filtering of the bottom-up signals will be consistent with the current emotional state, leading to what can be called the tunnel bias effect, helping our mind focus upon the most useful content in the moment.

ATTACHMENT D: GLOSSARY

Addiction—compulsive engagement with rewarding stimuli which continues despite negative consequences. *See also* Use and Misuse.

Affect—our muscle movements from emotions observed by another person, such as facial expressions, tone-of-voice, posture, hand gestures, and other physical movements. Affect is thought of as our emotional "weather."

Bias—an outcome from the emotional system functioning effortlessly and automatically without any deliberate controls swaying our opinions, judgments, choices, attitudes, and decisions. For more information see Attachment C: Biasing–A Few Concepts to Consider.

Bits and Pieces—how long-term memory is formed and stored as the important content of an experience gets captured for potential future use. The experience is organized from general content down to specific key elements in separate places (like visual features in one place, audio parts in another) linked together. It is consolidated and stored as a pattern of neural circuit connections now supported as long-term memory capable of lasting for hours, days, weeks, months, up to years.

Brain Plasticity—our brain's capacity to change based on lived experience by strengthening or weakening connections between neurons called the synapse. Lived experiences shape the memory needed to adapt to our environment helping us survive.

Chain of Bias Transmission—the process moving good, neutral, fabricated, or toxic ideas from person to person one link at a time. The legacy of the ideas gets so deeply rooted they continue to linger within a chain of bias transmission long after their usefulness. In order to slow down or stop the spread of out-of-date toxic ideas, methods to disrupt the chain of transmission needs to be found. But interrupting this bias chain is not so easy to do.

Concept—a small unit of knowledge stored in memory. Then, as we encounter various contexts in living our lives, the concept is used to classify or categorize the experience in some way, such as a negative threat, a positive opportunity or a neutral situation.

Conflict—the struggle due to opposing or incompatible interests, drives, wishes flowing from internal or external demands. Conflict is simply part of living life. There are a wide variety of reasons for conflicts forming in our mind as we sense something is wrong, such as a desire to go do something versus meeting the responsibilities at home or work.

Disinformation—an intentional effort to sow confusion by telling lies, or promoting untruths in order to gain influence or money.

Dispute—disagreement due to conflict, when one side (1) places blame on another, (2) makes direct demands, (3) then takes some form of action. The action may range from calling the question in a debate, a mild difference between those involved with talks taking place, an organized, legal claim filed in court, up to violent struggles between rival groups, including overt acts of warfare between nations. In most cases, talks take place with each side trying to make their case. If any reasonable solution is to be found, plenty of active listening by both sides is required.

Empathy—the simple idea of unconsciously sensing the emotions of another person.

Emotional Brains that Think—emotions serve an essential role in decision-making by giving meaning in the moment used to bias the options available before choosing. The importance of these biasing features can lead to the conclusion we have emotional brains that think.

Emotional Episode—the transition of one emotional state into another as the threatening stimulus or opportunity remains present. Basically, the emotional states keep flowing one into another as the situation unfolds.

Emotional Intelligence—the capacity to recognize, manage, and improve our own emotions as well as learn to notice emotions in other people. A good starting point in learning this skill is improving our self-awareness when experiencing negative or positive emotions.

Emotional State—an unconscious biological chain reaction measured in seconds to minutes, using neural circuits plus the bloodstream moving a variety of biochemical blends around the body to engage in short-term survival efforts, involving negative threats or positive opportunities.

Emotional Support—a micro level person to person effort starting as simple practical assistance. It can then involve listening to the other person's story in a safe, non-judgmental way, as well as other forms assistance.

Emotional System Outcomes—the primary result of the emotional system is successfully staying alive, influencing (1) Attention, (2) Long-term Memory, (3) Muscle Movements, and (4) Decision-making.

Emotional System Processes—For teaching purposes, our biological body can be divided into the various layers to describe the emotional system processes: the Foundation Part is the non-verbal and unconscious biological support needed for our body to function. The Foundation Part contains the Gut Reaction layer and Life Happens layer. The Upper Part is verbal thinking which provides our sense of mental actions, including Quick Thinking and Go Slow Thinking. These process layers are seamlessly blended together and fully connected to each other.

Emotions—natural, universal, biological functions needed to survive in the world while giving meaning to the experiences in life. They begin as an unconscious automatic appraisal to start survival-oriented actions, which in turn influence bodily functions, attention, thoughts, memories, feelings, and behaviors. Emotions work in the present to help shape the significance from our experiences and influence decisions on what to do next.

Engram—a pattern of brain connections linking together the bits and pieces of the content being stored in long-term memory.

Epigenetics—the impact of the environment influencing gene expression based on one's lived experience.

Explicit—something fully revealed and perfectly clear in its meaning. As a biological human feature, explicit functions usually suggest the focus of interest reaches our awareness, setting up the opportunity for a verbal description.

Explicit Bias—conscious awareness of verbally saying the words and/or taking the actions supporting our point of view.

Explicit Memory—a long-term memory function also referred to as declarative memory. Explicit declarative memories are stored via the hippocampus. They include the conscious memory of emotional experiences used to verbally recall the events, such as our personal history of places, people, and daily events of everyday life, giving us the ability to verbally recall experiences.

Explicit Teaching—the deliberate work of presenting specific content in a classroom or related learning platform. It is also referred to as explicit conditioning.

Feedback—a self-regulating mechanism used to monitor the failures and successes of the efforts made. Feedback is there to help support, regulate, adjust, and balance the system. It provides very important information to the emotional system on what is working and what is not, then helping to make the adjustments as needed. We need the feedback in order to make the necessary adjustments in an ever-changing world.

Feeling Emotions—sensing the biological changes inside our body during an emotional state. Feelings are the perceptions of the actual biological patterns, supporting the emotional state. Feelings range from nonconscious sensing up to full conscious awareness with the ability to use words to identify the emotion.

Fight—a mobilization strategy involving engagement with the threat because our life is in imminent danger. We need to act with aggression.

Flight—a mobilization strategy to escape from the threat, using the safest path possible.

Forget—a long-term memory function where over time, the stored but unused bits and pieces of memory are lost along the way.

Form—a long-term memory function consolidating short-term memory with emotional significance into bits and pieces for long-term storage.

Foundation Part—the nonverbal and unconscious biological support needed for our body to function. Think about this as an unconscious biological structure like a foundation below ground level, holding up and supporting a building. For teaching purposes, the foundation part has two layers: gut reaction and life happens.

Freeze—an immobilization strategy involving the shutdown and disengagement of the body. One does not move because predators are better at noticing moving prey.

Go Slow Thinking—verbal, conscious deliberate reasoning with effort. This upper part layer has the capacity to make conscious choices, providing the basis for self-monitoring. The Go Slow Thinking effort is needed to counter what can feel like the instinctive suggestion from Quick Thinking, if it is not too late. The Go Slow Thinking is the place where we consciously notice what is going on around us, making it the home of our adult in charge. It is the sunlight zone where considered, measured, deliberate, slower effort-needed thinking with strain is done. Go Slow Thinking is responsible for sorting out truth from fiction. The verbal feature may be speech or thoughts inside our head.

Gotcha Moments—deliberate efforts made by one to find and share unwarranted feedback. The one is focused on looking for something which can appear wrong, disturbing, or embarrassing to an audience. When found, the one proclaims "got you" or gotcha! Once caught, the action is quickly shared with the target individual and others.

Gut Reaction—a nonverbal and unconscious process layer, universally shared common biological blueprint hardwired into every person. It is based on the reality that all humans are 99.9 percent identical in our genetic makeup. *See also* Visceral Reaction.

Habits of Mind—unconscious patterns of thought providing a way of processing information leading to routine ways of thinking shaped by experiences and reality. These are the mindsets we use to successfully deal with various pressures in our life. Habits of mind are a blend of Quick Thinking and Go Slow Thinking, shaped over time. In the moment, they are influenced by emotions. Habits of mind are unique to the individual.

Homeostasis—the moment-to-moment process of the body, maintaining the chemical balances needed to meet the requirements of keeping stable internal biological resources under ever-changing conditions. The internal balancing includes maintaining the right temperature, blood levels for certain chemicals, demands by the body for resources like oxygen, water, specific foods, sleep, and other requirements. This balance is needed for our body to respond to the demands of living and maintaining health, as well as any needed repairs.

Hormones—biochemicals with names like adrenaline, cortisol, and oxytocin. They are produced by the body's endocrine glands and, when released, flow through the bloodstream, influencing the target cells, tissues, and organs. . They are part of the body's internal complex and nuanced messaging methods influencing our different responses to life events.

Human-Caused Dangerous Events—experiences involving military combat, mass fatality shooting, sexual assault, childhood abuse, domestic violence, intimidation, physical abuse, self-inflicted wounds, and other related events.

Implicit—implied and unexpressed influences. As a biological human feature, implicit functions are in the foundation part, operating unconsciously, automatically, and reflexively. There are a variety of implicit influences never reaching our awareness.

Implicit Bias—the effortless automatic emotional part working deeply hidden below conscious awareness without any deliberate controls. All of us have implicit bias in the moment as a natural inclination, not obvious or apparent to us, unconsciously swaying our opinions, judgments, choices, attitudes, and decisions.

Implicit Memory—a long-term memory function also referred to as nondeclarative memory. The implicit nondeclarative memory is managed by the amygdala, operating automatically and reflexively, providing unconscious recall that never reaches our awareness. We are unaware of implicit emotional memory recall, but it does influence our actions. It includes procedural memory, habituation (learning to ignore neutral, nonthreatening stimuli), sensitization (having a stronger reflexive response to certain stimuli) priming cues, puzzle-solving skills, classical conditioning, and building associations between memories.

Implicit Teaching Methods—implied, wordless lessons demonstrated over time by leaders, role models, seasoned group members, and in other unexpressed ways. The students are expected to imitate the unsaid implied details. We are not conscious of these implicit lessons; however, over time, they do form, mold, and shape our thinking. This in turn influences our actions and behavior. Many cultural features are taught, using implicit means. Then we get good at what we practice with the full support by those around us. It is also referred to as implicit conditioning.

Input—in business, inputs include the raw materials, facility space, equipment, supplies, technology, electricity, gas, trained workers, administrative structures (like policies, procedures, contracts), and related items as needed to engage in the value-added process. For a biological body, input includes not only food, water, oxygen to breathe, but also incoming stimuli received through the outward-looking senses (sight, sound, touch, taste, smell), and internal thoughts in our head (something recalled or imagined). These inputs serve as a trigger to start an unconscious automatic appraisal, using biological functions.

Interoception—the perception of feeling the process of balancing the body's chemical brew by reading the viscera and homeostasis. The brain is monitoring homeostatic changes in the viscera (internal organs like the heart, lungs, and stomach), hormones flowing in the bloodstream, immune system activity, and related biological functioning as needed. Our brain simulates our current body state to make predictions on resources needed, like food and water. Then to balance this chemical brew, we experience a range of sensations like cravings and urges to motivate actions, ranging from unpleasant to pleasant. The term interoception refers to how the functioning of our internal organs, hormones flowing in our blood steam, tissues, immune system are represented in our brain, influencing this balancing activity. Interoception is a core part of emotions.

Interests—needs we want to satisfy, such as food, clothing, and shelter; the desire to be treated with respect, be understood, have control over our life, privacy, safety, financial security, or trust. Interests may be based on principles, beliefs, and values. It might involve control over material goods, reputation, job responsibilities, payment due for services delivered, liability for damaged property, parenting time with the children, and much more.

Issues—problems or topics we consider basic to the dispute based on the interests.

Life Happens—a nonverbal and unconscious process layer unique to each individual. It starts with the 0.1 percent genetic makeup specific to each individual. It is further shaped by the impact of the environment where the person lives via epigenetics and brain plasticity.

Long-Term Memory—forming, storing, recalling, and forgetting experiences. Emotional memories formed and stored in the past are recalled, influencing actions right now in the present. Content not used is forgotten.

Macro Level—involves large groups working within the same culture. It is the place for intergroup relationships connected together by a shared culture, using the same jargon. In between the universal all and unique one are features groups of people share in common, such as language, units of measurement, and values.

Meta Level—where the decisions get made to organize overall society. What cannot be managed at the macro level gets sorted out at the meta level. This is where the work of addressing competing interests happens as the political will of the top officials shapes social, public, and institutional reality for the whole of the people.

Micro Level—is between individuals, as in people who can communicate face-to-face. It is the place for the interpersonal, small intragroup relationships we have at home, work, and other social settings.

Misinformation—an unintentional misinterpretation of events with the mistaken content shared with others. As part of the fog of uncertainty mistakes get made. Then as ideas are exchanged and suitable feedback is received, the misinformation gets sorted out with corrections made.

Misuse—when the use engagement gets extended and it is harder to stop. Not all positive rewarding stimuli are good things. *See also* Use and Addiction.

Moods—persistent emotions coloring the perception of one's life remaining consistent for extended time periods, measured in minutes, hours, or days. Because they are sustained and pervasive, moods are considered to be one's emotional "climate."

Nano—A nanometer is the tiniest unit of measurement at one billionth part of the whole. Nano biological functions start at the very smallest places inside the cell, driven by the universal features of our genetics and spread across our entire body, including the brain. For example, according to the National Nanotechnology Initiative, a strand of human deoxyribonucleic acid (DNA) is 2.5 nanometers in diameter. Thus, emotions start as molecular, microscopic, biological functions measured in nanometers with the task of looking out for our own self-interest to survive.

Nano Level—intrapersonal playing out inside our mind and body, including self-talk.

Natural Disasters—dangerous life-threatening experiences including fires, tornadoes, floods, earthquakes, hurricanes, and other related events as well as sudden medical issues.

Negative Emotions— more intense, narrowed, and focused efforts in response to threats.

Neuroception—a term coined by Stephen W. Porges. It means our brain is hard-wired to monitor for threats and avoid danger. This sets up a natural universal negative bias needed for the continuous, unconscious monitoring of dangers and threats. Behaviorally, our guard is up as negative emotions serve as the default setting.

Neurons—specialized cells with the capacity to influence other cells. Where two neurons meet, there is a very small gap called the synapse, measured in nanometers. An electrical impulse traveling inside the neuron triggers the release of the chemical signal, sending it across the gap to the next nerve cell. Neurons linked together create circuits to communicate, using a blend of electricity as well as these chemical messengers called neurotransmitters.

Neurotransmitters—a specialized biochemical agent released into the synapse triggered by the electrical impulse traveling inside a neuron. Neurotransmitters, with names like dopamine and serotonin, are part of the body's internal complex, nuanced messaging methods that influence our different responses to life events.

Objective—a desired outcome achieved by making the effort to reduce distortion from emotions. Commonly referred to as flowing from our "head" objective suggests dealing with conditions perceived as being without influence from emotions, feelings, bias, or prejudices. Facts can then be established using deliberate methods to show something actually physically exists in reality. These efforts work to reduce subjective influences to a minimum.

Output—the final end results and outcomes from the value-added processes in a system.

Physical Elements—the composition, interactions, processes, and known properties of what is around us and happens in fact. Physical elements do not change. Human understanding of these physical elements does change by using the scientific method. For example, science established the reality of how the Earth orbits the Sun while rotating.

Positions—specific proposals or demands, framed as solutions to resolve the issues based on the interests.

Positive Emotions—widen perspective and add to useful resources in response to opportunity.

Priming—exposure to novel, new, never seen, heard, smelled, tasted, or touched before content is placed in our memory as a first-time experience. Once established, the next similar experience will be linked by association to the first exposure. Some call this the mere exposure effect.

Procedural Memory—sometimes referred to as muscle memory, is needed to learn skills involving automatic repeated movements. It is created by practicing the necessary steps in the right order over many repetitions. These repeated actions create, strengthen, and help to condition the neural circuits needed to support this memory. The neural circuits are shaped by these repeated exposures leading to an accumulation of lots of small changes and tiny improvements. Over time, this aggregation of small changes has an impact. Regular practice then continues to improve and maintain this skill. Then procedural memory allows us to do the appropriate muscle movement when it matters accomplishing the results we were seeking.

Process—a series of vital tasks performed in a sequence, leading to results. For teaching purposes, our biological body can be divided into layers of emotional processes. These layers are seamlessly blended together and fully connected.

Quick Thinking—verbal, automatic, fast, instinctive, reacting first without considering the actions done or the words spoken. It is non-conscious because the words can be spoken without much thought or awareness. Quick Thinking can seem like we are cruising with the ease of autopilot. This includes the feeling of emotions while not being consciously aware of them. The verbal feature may be spoken or thoughts inside our head.

Recall—a long-term memory function where cues trigger retrieval of some bits and pieces assembled in the moment for use now.

Rule of Repetition—saying words and/or doing actions in a consistent manner multiple times as needed is essential to communicate content. In messaging, we cannot expect to say or do something once and have the target audience correctly remember the content, based on how attention and long-term memory operate.

Short-term Memory—brain structures used for attention and focus, including an appraisal of emotional meaning, which can provide a quick response to a dangerous threat or a fine opportunity. Then selected parts of what we have heard, seen, tasted, smelled, touched, or thought in short-term memory go through a consolidation process to form and store the content for later use. Significant emotional moments are given a higher priority.

Social Features—a result of the human mind interpreting physical elements. It includes people agreeing on the words used to describe and give meaning to the physical facts. From these interpretations, we learn concepts. For example, social features created the designation of time it takes the Earth to orbit the Sun as one year and the Earth completing a rotation as one day.

Store—a long-term memory function where an engram is set up to hold the content of the experience for a while ranging from a few minutes up to many years.

Stress—the natural biological response to external challenges and threats experienced as part of meeting the demands of daily living. Our perception of the pressure, strain, or tension is important to reading the situation. In response, our body ramps up the resources needed to immediately address the threat, such as triggering the sympathetic nervous system's fight or flight. Stress can be positive or negative. Challenge stress does act as a motivational force. The source of the threat or challenge may range from a one-time, short-term event up to something happening repeatedly over an extended period of time. The impact of chronic accumulative stress can lead to burnout, which is feeling physically exhausted and emotionally drained due to this prolonged pressure, strain, or tension. Stress does need to be managed as part of the way we properly maintain our body. If we do not, there will be problems arising from this basic wear and tear on our body due to the constant demands.

Subjective—emotions influencing preferences. Commonly referred to as flowing from our "heart." Subjective suggests using positive and negative emotions to impact how we evaluate options leading to favoring particular ingredients, features, or other qualities to form opinions and make judgments. These preferences are not necessarily based on reality or having substance.

System of Prejudice—within a culture, social practices formed over a long period of time imposed by a dominant group created with ex-plicit intent to manage a targeted subordinate group. The dominant leaders can choose to establish social practices based on biological features, economic status, political affiliation, geographic location, religious preferences, medical impairments, or anything else they select. The people on the receiving end of these social practices ex-perience systemic racism, sexism, or other forms of discrimination.

System Theory—a business management tool used in quality im-provement to look at input, process, output, feedback.

Traumatic Memory—stored memory from directly experiencing or witnessing a dangerous life-threatening event. The negative emotion-al event is formed and stored as a blend of implicit and explicit long-term memory. Then, in the present moment, triggers cause parts of the past traumatic incident to be recalled, experienced as flashbacks. The reality with trauma is the scars in our mind long outlive the physical wounds healing. The emotional memory continues to play out in our body, actions, and thinking.

Trigger—a cue or signal prompting action in the form of an emotional state.

Unintended Human Failure—Experiences involving injury and death from a serious accident, technology tragedy such as a structural collapse, and other related events.

Unique One—the 0.1 percent genetic makeup specific to each individual, distinctive enough to be used as evidence in a court of law to establish presence or absence at a crime scene. It is further shaped by the impact of the environment where the person lives. Epigenetics and brain plasticity provides lived experience shaping the memory needed to adapt to our environment, helping us survive.

Universal All—based on the fact that all humans are 99.9 percent identical in our genetic makeup. Thus, deep inside each core human body, we are all the same.

Upper Part—Quick Thinking and Go Slow Thinking are two layers of conscious mental actions like one is upstairs in the living area of the building as seen from the street. Here, our thinking provides a sense of mental actions which can be verbally expressed. The upper part includes routine ways of thinking called habits of mind as influenced by our Quick Thinking and Go Slow Layers. The verbal feature may be spoken or thoughts inside our head.

Use—involves engagement with rewarding stimuli, like good food, quenching our thirst, and positive social interactions, all leading to the enjoyable experience we want to repeat again. *See also* Misuse and Addiction.

Us-versus-Them—a line of thinking becoming part of our habits of mind at both the micro and macro levels. The "us" feature involves identifying with our own group, based on some unifying theme like geographic area, shared purpose (mission, vision, values), vocation, age, sex, skin color, cultural interests, political issues, and more. The other group has some clearly distinguishing features, such as wearing the rival team colors, a business competitor, up to and including political opponents. The "them" feature involves the perception of an imminent danger, a challenge, something at stake, framed as a threat to our group's very existence. In the extreme, the leader's messaging is filled with polarizing rhetoric lifting up "us" and demonizing "them" because the threat remains and the clock is ticking.

Vagus Nerve—the longest cranial nerve, extending into the chest and abdomen with functions including swallowing, muscles of the vocal cords, heart rate, breathing, and gastrointestinal tract (gut).

Visceral Reaction—a core feature of the gut reaction. A visceral reaction is sensing biological changes to the internal organs like heart, lungs, stomach, and gut as part of the bidirectional connections between the insular cortex and vagus nerve communicating with core areas of the body. There is an insular cortex on the left and right side of the brain, close to its center. The insula cortices provide an integrative role for information flow across the body from the internal organs up into the brain on how these tissues are functioning. The insula sends out instructions to the visceral organs like: heart beat faster, lungs breath more rapidly, or pause digestion for now. These are the biological changes happening during an emotional state experienced as feelings. *See also* Gut Reaction.

INDEX

ABOUT THE AUTHOR

James Scott Harvey

Jim has over forty-six years of human services experience, working in the areas of behavioral health, mediation, and more. His work has included direct micro services, administrative macro services, and teaching. He used his lifetime of experiences to develop the ideas inside this book. He has a Master of Social Work degree. Each day, he finds a few minutes to play ukulele.

ACKNOWLEDGMENTS

This work would not have been possible without the love and support from my wife Maureen.

To my brother Jeff who patiently listened to my efforts to develop this book and provided many useful suggestions along the way.

To John Wackman who was my dear lifetime friend. He encouraged my early efforts in developing this work as well as the repair of beloved but broken items.

Thank you to Robert Garnett who provided useful comments early in the beta review.

To Kathryn Bellman who read and provided valuable feedback at a key moment in this work.

To David Hubbard, Casey Karges, Sovida Tran, Mary Brunning, Lisa Pytlik Zillig, plus the many other fine people associated with the mediation community who provided plenty of encouragement in developing this project over many years.

ENDNOTES

1   National Institute of Neurological Disorders and Stroke (2013) "Brain Basics: Know Your Brain." National Institutes of Health (Office of Communications and Public Liaison). NIH Publication No.11– 3440a. Last updated December 19, 2013. Accessed on January 30, 2014 http://www.ninds.nih.gov/disorders/brain_basics/know_your_brain.htm#art

2   *THE HUMAN BRAIN IS THE MOST COMPLEX STRUCTURE IN THE KNOWN UNIVERSE.*
- Iacoboni (2008) *Mirroring People;* page 8
- Klingberg (2009). The Overflowing Brain. page 125.

3   Most of our brain at work is unconscious.
- Ayan (2015) Eric R. Kandel discusses Freud's legacy, memory's foibles and the potential of drugs that boost brainpower. Scientific American Innovators: Memory. Topix Media Lab Special #2, 2015. page 23.
- Eagleman (2015) The Brain The Story of You. page 97.
- Sapolsky (2016) We're rarely rational when we vote because we're rarely rational, period. Los Angeles Times. Op-Ed \ Opinion. April 3, 2016. Accessed on April 7, 2016. http://www.latimes.com/opinion/op-ed/la-oe-sapolsky-how-we-decide-how-to-vote-20160403-story.html
- Damasio (2010) on Self Comes to Mind—Number Six in a Series. accessed on April 25, 2017.

Uploaded on Nov 10, 2010. https://www.youtube.com/watch?v=Aw2yaozi0Gg

4   Shakespeare (1599–1602) The Tragedy of Hamlet, Prince of Denmark (Act 2, Scene 2, line 246).

5   All Conflict Starts and Ends in the Brain, Driven by Emotions
- Mental Health: A Report of the Surgeon General (1999) National Institute of Mental Health. Chapter 2—"The Fundamentals of Mental Health and Mental Illness." accessed on July 29, 2021 https://profiles.nlm.nih.gov/spotlight/nn/catalog/nlm:nlmuid-101584932X120-doc
- Kandel (2006) In Search of Memory: The Emergence of a New Science of Mind.

6   Concepts
- Feldman Barrett (2017) How Emotions Are Made.

- Caldwell-Harris (2019) Our Language Affects What We See. January 15, 2019. Scientific American Mind & Brain (email) January 16, 2019. Accessed on January 20, 2019. https://www.scientificamerican.com/article/our-language-affects-what-we- see/?utm_source=newsletter&utm_medium=email&utm_ campaign=mind&utm_content=link& utm_term=2019–01–16_featured-this-week&spMailingID=58248795&spUserID=MTUwMTI2MDc0NDY4S0&sp JobID=1562061894&spR eportId=MTU2MjA2MTg5NAS2

[7] Positive and Negative emotions.
- Fredrickson (2003) "The Value of Positive Emotions. The Scientific Research Society. American Scientist. 2003 July–August.
- Feldman Barrett (2007) The Science of Emotion. National Research Council. 2008. Human Behavior in Military Contexts. https://doi.org/10.17226/12023. Accessed on June 1, 2014.
- Fredrickson (2009) Positivity: Top-Notch Research Reveals the 3 to 1 Ratio That Will Change Your Life.
- Davidson, Begley (2012) The Emotional Life of Your Brain.

[8] Living Single Cell Organisms
- Damasio (1999) The Feeling of What Happens;
- Damasio (2003) Looking for Spinoza: Joy, Sorrow, & the Feeling Brain;
- Kandel (2006) In Search of Memory. page xiii
- Damasio (2010) Self Comes to Mind: Constructing the Conscious Brain.

[9] Emotional State
- Feldman Barrett (2017) How Emotions Are Made.
- Damasio (2010) Self Comes to Mind.
- Davidson, Begley (2012) The Emotional Life of Your Brain.
- Taylor (2006) My Stroke of Insight.
- Hochman (2020) "A Rebooted Brain—Jill Bolte Taylor, 60. AARP The Magazine. Feb/ Mar 2020.
- Ratey (2001) A User's Guide to the Brain: Perception, Attention, & Four Theaters of the Brain.

[10] Biochemical blends (many sources including)
- The Society for Endocrinology (2021) You and Your Hormones, Glossary. Accessed on August 3, 2021. https://www.yourhormones.info/glossary/a#
- Sapolsky (2017) Behave. The Biology of Humans at Our Best and Worst. Appendix 2—the Basics of Endocrinology.
- The Whole Psychiatry & Brain Recovery Center (2021) Neurotransmitters Table. accessed on August 3, 2021. https://wholepsychiatry.com/physician-resources/neurotransmitters-table/
- Carter (2009) The Human Brain Book.

[11] Affect
- Damasio (1999) The Feeling of What Happens.
- Baker and Trzepacz (1998) Chapter 2—Mental Status Examination. Koocher, Norcross, Hill (Editors) (1999) Psychologists' Desk Reference.
- American Psychiatric Association (2013) Diagnostic and Statistical Manual of Mental Disorders— Fifth Edition (DSM-5).
- Merriam-Webster's 11th Collegiate Dictionary.

12 Mood
- Damasio (1999) The Feeling of What Happens.
- Damasio (2003) Looking for Spinoza.
- Davidson, Begley (2012) The Emotional Life of Your Brain.
- Levitin (2006) This Is Your Brain on Music.
- American Psychiatric Association (2013) Diagnostic and Statistical Manual of Mental Disorders— Fifth Edition (DSM-5).

13 Feelings of Emotions
- Damasio (1999) The Feeling of What Happens.
- Damasio (2003) Looking for Spinoza.

14 Bias—In this context the word *bias* suggests partiality, unfairness, or favoritism.
- Banaji and Greenwald (2013) BLIND SPOT: Hidden Biases of Good People.
- Merriam-Webster's 11th Collegiate Dictionary

15 Neuroception
- Porges (2007). The Polyvagal Perspective. Biological Psychology, 74(2), 116–143. Accessed on July 30, 2015. http://www.ncbi.nlm.nih.gov/pmc/articles/PMC1868418/
- Porges (2011) The Polyvagal Theory.
- van der Kolk (2014) The Body Keeps the Score.

16 Trauma:
- Substance Abuse and Mental Health Services Administration (2014) Concept of Trauma and Guidance for a Trauma-Informed Approach. https://store.samhsa.gov/shin/content/SMA14– 4884/SMA14–4884.pdf. Accessed Sept 16, 2017
- Substance Abuse and Mental Health Services Administration (2017) Resource Guide to Trauma- Informed Human Services. https://www.acf.hhs.gov/trauma-toolkit. Accessed Sept 15, 2017
- Substance Abuse and Mental Health Services Administration (2015) Trauma-Informed Care in Behavioral Health Services. Accessed on September 28, 2020. https://store.samhsa.gov/product/Trauma-Informed-Care-in-Behavioral-Health-Services/SMA15–4420
- American Psychiatric Association (2013) Diagnostic and Statistical Manual of Mental Disorders— Fifth Edition (DSM-5). acute stress disorder (308.3), posttraumatic stress disorder (309.81).
- Institute of Medicine. 2003. Preparing for the Psychological Consequences of Terrorism. Chapter 2. Accessed April 26, 2019. https://www.nap.edu/read/10717/chapter/4
- Schiraldi (2009) The Post-Traumatic Stress Disorder Sourcebook. Table 1.1

17 Examples of Language used to Label Emotional Concepts
- Ekman, Friesen (2003) Unmasking the Face.
- Ekman (2004) Emotions Revealed.
- Davidson, Begley (2012) The Emotional Life of Your Brain.
- Reilly & Shopshire (2019) Anger Management for Substance Use Disorder and Mental Health Clients. Substance Abuse and Mental Health Services Administration.
- Kuppens, Allen, and Sheeber (2010). Emotional inertia and psychological

maladjustment. Published online 2010 May 25. http://www.ncbi.nlm.nih.gov/pmc/articles/PMC2901421/ accessed on July 30, 2020.
- American Psychiatric Association (2013) Diagnostic and Statistical Manual of Mental Disorders— Fifth Edition (DSM-5).
- Fredrickson (2009) Positivity: Top-Notch Research Reveals the 3 to 1 Ratio That Will Change Your Life.
- Merriam-Webster's 11th Collegiate Dictionary
- American Heritage College Dictionary

[18] System
- Merriam-Webster's 11th Collegiate Dictionary.
- Scholtes (1998) The Leader's Handbook—Making Things Happen, Getting Things Done.
- French and Bell (1973) ORGANIZATION DEVELOPMENT Behavioral Science Interventions for Organization Improvement.

[19] Emotional Process—Foundation Part and Upper Part
- Mlodinow (2012) Subliminal: How Your Unconscious Mind Rules Your Behavior.
- Discussing William Benjamin Carpenter (1874) two trains concept.
- LeDoux (1996) The Emotional Brain: The Mysterious Underpinnings of Emotional Life.
- LeDoux (2002) Synaptic Self: How Our Brains Become Who We Are.
- LeDoux (2006) The Power of Emotions. The Dana Foundation, New York, NY. http://www.dana.org/Publications/ReportDetails.aspx?id=44196. Accessed on May 14, 2015.
- Goleman (2006) Social Intelligence: The New Science of Human Relationships.
- Seubert and Regenbogen (March 2012) "I Know How You Feel: Good social skills depend on picking up on other people's moods" Scientific American Mind.
- Damasio (2003) "Looking for Spinoza: Joy, Sorrow, and the Feeling Brain"
- Weintraub (2012) Discover Interview: Jaak Panksepp. Discover Magazine (from the May 2012 issue). Accessed on March 24, 2015.
- http://discovermagazine.com/2012/may/11-jaak-panksepp-rat-tickler-found-humans-7-primal-emotions

[20] All Brains Have Common Structures and Yet Each Brain Is Unique.
~ALL~ Global Universal ~ Common Blueprint
- National Human Genome Research Institute (NHGRI) Accessed on January 16, 2014
- http://www.genome.gov/
- Deoxyribonucleic Acid (DNA) Fact Sheet. accessed on December 6, 2020.
- Deoxyribonucleic Acid (DNA) Fact Sheet (genome.gov)
- Whole Genome Association Studies. Last Reviewed: July 15, 2011. National Human Genome Research Institute, National Institutes of Health. http://www.genome.gov/17516714 Accessed on January 8, 2015; and December 6, 2020.
- When researchers completed the final analysis of the Human Genome Project in April 2003, they confirmed that the 3 billion base pairs of genetic

letters in humans were 99.9 percent identical in every person. It also meant that individuals are, on average, 0.1 percent different genetically from every other person on the planet.

- National Institutes of Health (NIH), National Human Genome Research Institute. Last updated: September 7, 2018. Accessed on December 6, 2020. https://www.genome.gov/19016904/faq-about-genetic-and-genomic-science/ "All human beings are 99.9 percent identical in their genetic makeup." is under the section "Why are genetics and genomics important to my health?"
- Wells (2006) Deep Ancestry: Inside the Genographic Project

[21] Homeostasis and Interoception
- Merriam-Webster's 11th Collegiate Dictionary
- Damasio (2010) Self Comes to Mind: Constructing the Conscious Brain.
- Feldman Barrett (2017) How Emotions Are Made: The Secret Life of the Brain.

[22] Epigenetics
- Higgins (2008) "The New Genetics of Mental Illness"; Scientific American Mind, June/July 2008.
- Gage and Muotri (2012) "Jumping Genes in the Brain Ensure That Even Identical Twins Are Different. Scientific American Magazine. March 2012
- Rodriguez (2015) Epigenetics: Descendants of Holocaust Survivors Have Altered Stress Hormones. Scientific American Mind, March/April 2015.
- Davidson, Begley (2012) The Emotional Life of Your Brain.
- National Institute of Mental Health (NIMH) "Brain Basics" accessed on January 17, 2014 http://www.nimh.nih.gov/health/educational-resources/brain-basics/brain-basics.shtml
- National Research Council and Institute of Medicine. (2009). Preventing Mental, Emotional, and Behavioral Disorders Among Young People. Glossary. Accessed on January 9, 2014. http://books.nap.edu/openbook.php?record_id=12480&page=R2
- Park (2012) Don't Trash These Genes. Time Magazine. October 22, 2012.
- Squire & Kandel (2000)—Memory: from Mind to Molecules.
- Alcock (2001) The Triumph of Sociobiology.

[23] Gleick (1987) Chaos: Making a New Science.
[24] The Science of Quick Thinking and Go Slow Layers
- Kahneman (2011) Thinking, Fast and Slow.
- Kahneman (2002) "Maps of Bounded Rationality: A Perspective on Intuitive Judgment and Choice" Prize Lecture, December 8, 2002—The Sveriges Riksbank Prize in Economic Sciences in Memory of Alfred Nobel 2002. Accessed on August 7, 2014. http://www.nobelprize.org/nobel_prizes/economic-sciences/laureates/2002/kahneman-facts.html
- Bargh (2014) Our Unconscious Mind: Unconscious impulses and desires impel what we think and do in ways Freud never dreamed of. Scientific American. January 2014.
- Myers (2007) The Powers and Perils of Intuition: Understanding the nature of our gut instincts.

- Scientific American Mind. June/July 2007. http://www.scientificamerican.com/article/powers-and-perils-of-intuition/
- Keltner, Oatley, Jenkins (2013) Understanding Emotions (Third Edition).

25 Nano level
- Merriam-Webster's 11th Collegiate Dictionary
- United States National Nanotechnology Initiative. Accessed on September 25, 2015. http://www.nano.gov/nanotech-101/what/nano-size.

26 Senses
- van der Kolk (2014) The Body Keeps the Score. page 70
- Stickgold (2015) Sleep on It! Scientific American. October 2015. Page 54
- Purves, Augustine, Fitzpatrick, Hall, LaMantia, McNamara, and Williams (Editors) (2004) Neuroscience: Third Edition. Unit II.
- Monell Chemical Senses Center (2020) WHAT IS OLFACTION? accessed on July 28, 2020 https://www.monell.org/research/anosmia/how_smell_works
- Carter (2009) The Human Brain Book. Page 60
- Fields (2009). The Other Brain. Page 321

27 Dobbs (2006) Master of Emotions (on Joesph LeDoux). Scientific American Mind. Feb-Mar 2006.

28 Additional second or two
- Purves, Augustine, Fitzpatrick, Hall, LaMantia, McNamara, and Williams (Editors) (2004). Neuroscience: Third Edition.
- McCracken, General Editor (2000) New Atlas of Human Anatomy
- Restak (2001) The Secret Life of the Brain.
- Ratey (2001) A User's Guide to the Brain: Perception, Attention, & Four Theaters of the Brain.

29 Habit of Mind—a blend of Quick and Go Slow Thinking
- Facione (2015) "Critical Thinking: What It is and Why it Counts." www.insightassessment.com. accessed on November 30, 2016.
- Costa (Dec 2008) Chapter 2. Describing the Habits of Mind. *Learning and Leading with Habits of Mind* Edited by Arthur L. Costa and Bena Kallick. http://www.ascd.org/publications/books/108008/chapters/Describing-the-Habits-of-Mind.aspx Accessed on September 14, 2015.
- Cather (Author), Jewell and Stout (Editors) (2013) Selected Letters of Willa Cather.

30 Attention Spotlight
- Damasio (2003) Looking for Spinoza: Joy, Sorrow, and the Feeling Brain.
- Piore (2016) Tuning Out the Noise: Neuroscientist and psychiatrist Michael Halassa, MD, PhD, is cracking the neurobiological puzzle of how we pay attention in a world of distractions. As reported in Brain in the News, The Dana Foundation. May 2016.

31 Attention system filters:
- Habituation & Sensitization: Squire and Kandel (2000) Memory: From Mind to Molecules.
- Novelty Detection: Kandel (2006) In Search of Memory.
- Reward Pathway: National Institute on Drug Abuse (2007) The Neurobiology of Drug Addiction; Section II: The Reward Pathway and Addiction. Accessed

on July 12, 2015. http://www.drugabuse.gov/publications/teaching-packets/
neurobiology-drug-addiction
- Emotional Memory
  o McGaugh, LePort (2015) Remembrance of All Things Past. Scientific
    American Innovators: Memory. Topix Media Lab Special #2, 2015.
  o McGaugh (2003) Memory and Emotion.
  o Schacter (2001) The Seven Sins of Memory.
- Self-Talk Scripts: Fernyhough (2017) Talking to Ourselves. Scientific
  American. August 2017.
- Bodily Needs (Body Budget): Feldman Barrett (2017) How Emotions Are
  Made.

[32] Attentional Blink & Inattention Blindness
- Klingberg (2009) The Overflowing Brain.
- Davidson, Begley (2012) The Emotional Life of Your Brain.
- Caldwell-Harris (2019) Our Language Affects What We See. Scientific
  American Mind & Brain <news@email.scientificamerican.com> January 16,
  2019. Accessed on January 20, 2019. https://www.scientificamerican.com/
  article/our-language-affects-what-we- see/?utm_source=newsletter&utm_
  medium=email&utm_campaign=mind&utm_content=link&utm_
  term=2019–01–16_featured- this-week&spMailingID=58248795&spUse
  rID=MTUwMTI2MDc0NDY4S0&spJobID=1562061894&spReportId=
  MTU2MjA2MTg5NAS2
- Nieuwenstein, Potter, and Theeuwes (2009) Unmasking the attentional
  blink. J Exp Psychol Hum Percept Perform. 2009 Feb; 35(1): 159–169.
  Accessed on January 21, 2019. https://www.ncbi.nlm.nih.gov/pmc/articles/
  PMC2632766/
- Strayer and Watson (2012) Top Multitaskers Help Explain How Brain Juggles
  Thoughts. Scientific American Mind. March/April 2012 Issue.

[33] Memory
- McGaugh (2003) Memory and Emotion.
- LeDoux (2007) Scholarpedia, http://www.scholarpedia.org/article/Emotional_
  memory. Accessed on February 25, 2014
- Trafton (2017) Neuroscientist Identify Brain Circuit Necessary for Memory
  Formation. MIT News Office (April 6, 2017). Reprint in "Brain in the News"
  Dana Foundation. Vol. 24 No. 5. May 2017
- Squire and Kandel (2000). Memory: From Mind to Molecules.
- Damasio (2010) Self Comes to Mind.
- Major Memory Systems | Declarative—Explicit Memory and Non-
  Declarative—Implicit Memory.
- Squire (November 2007) Learning and Memory—The Dana Guide http://
  www.dana.org/news/brainhealth/detail.aspx?id=10020. Accessed on January
  11, 2015.
- Squire and Kandel (2000) Memory: From Mind to Molecules.
- Klingberg (2009). The Overflowing Brain.
- Purves, Augustine, Fitzpatrick, Hall, LaMantia, McNamara, and Williams
  (Editors) (2004) Neuroscience: Third Edition. page 734, glossary G-4.

- McGaugh (2003) Memory and Emotion.
- Kandel (2006), In Search of Memory.
- Richmond, Nelson (2007) Accounting for change in declarative memory. Developmental Cognitive Neuroscience. Volume 27, Issue 3, September 2007 http://www.ncbi.nlm.nih.gov/pmc/articles/PMC2094108/figure/F2/ accessed on January 30, 2016
- Sapolsky (2003) Taming Stress. Scientific American, September 2003.

[34] Engram—a pattern of connections
- Shen (2018) Portrait of a Memory. Nature magazine. March 14, 2018. Accessed via email from: Scientific American MIND Date: April 13, 2018. Subject: May/June Issue: The Science of Memory (Scientific American Mind, Volume 29, Issue 3). The article was reproduced with permission and was first published on January 10, 2018. https://www.scientificamerican.com/article/portrait- of-a-memory/?utm_source=promotion&utm_medium=email&utm_campaign=minddigital- may2018-issuealert-&utm_content=art   OptinNo
- Chodosh (2016) How the Brain Builds Memory Chains. Scientific American Weekly Review (email) Date: July 27, 2016. Accessed on 7/28/2016. http://www.scientificamerican.com/article/how-the-brain-builds-memory-chains/?WT.mc_id=SA_WR_20160727
- Gabrielsen (2016) The Cobblestones of Memory Lane. Proto Magazine. November 3, 2016. As printed in Brain in the News. January 2017. Vol. 24, No. 1.
- Purves, Augustine, Fitzpatrick, Hall, LaMantia, McNamara, and Williams (Editors) (2004) Neuroscience: Third Edition. Glossary page G-5
- Schacter (1996) Searching for Memory.
- Tonegawa, et al. (2015) Memory Engram Cells Have Come of Age. Neuron , September 2015. http://www.cell.com/neuron/abstract/S0896-6273(15)00677-7. Accessed May 12, 2017.

[35] Long-term memory organized.
- Quiroga, Fried, and Koch. Brain Cells for Grandmother. Scientific American Innovators: Memory. Topix Media Lab Special #2, 2015.
- Tsien (2015) The Memory Code. Scientific American. Innovators: Memory. Topix Media Lab Special #2, 2015.
- Schacter (2001) The Seven Sins of Memory.
- Makin (2017) Where Does the Brain Store Long-Ago Memories? Scientific American (Neuroscience) April 12, 2017. Accessed on April 19, 2017 via email from Scientific American Mind & Brain https://www.scientificamerican.com/article/where-does-the-brain-store-long-ago- memories/?WT.mc_id=SA_MB_20170419
- Memory
- McGaugh (2003) Memory and Emotion.
- LeDoux (2007), Scholarpedia, http://www.scholarpedia.org/article/Emotional_memory accessed on February 25, 2014.
- Trafton (2017) Neuroscientist Identify Brain Circuit Necessary for Memory Formation. MIT News Office (April 6, 2017). Reprint in "Brain in the News" Dana Foundation. Vol. 24 No. 5. May 2017.

- Squire and Kandel (2000). Memory: From Mind to Molecules.
- Damasio (2010) Self Comes to Mind.

36 Memory challenges
- Schacter (2001) The Seven Sins of Memory.
- DK (2017) The Psychology Book. pages 208–209.

37 Feldman Barrett (2017) How Emotions Are Made.

38 Affect
- Damasio (1999) The Feeling of What Happens.
- Baker and Trzepacz (1998) Chapter 2—Mental Status Examination. Koocher, Norcross, Hill (Editors) (1999) Psychologists' Desk Reference.
- American Psychiatric Association (2013) Diagnostic and Statistical Manual of Mental Disorders— Fifth Edition (DSM-5).
- Merriam-Webster's 11th Collegiate Dictionary.

39 Facial expressions read by our brain.
- Tsao (2019) Face Values. Scientific American. February 2019. Vol 320, Number 2.
- Tsao and Livingstone (2008) Mechanisms of face perception. *Annual review of neuroscience, 31,* 411–37 https://www.ncbi.nlm.nih.gov/pmc/articles/PMC2629401/ Accessed on February 15, 2019.
- Mlodinow (2012) Subliminal: How Your Unconscious Mind Rules Your Behavior. page 38.

40 Prosopagnosia
- Merriam-Webster's 11th Collegiate Dictionary.
- Nietzell, Bernstein, Milich (1994) Introduction to Clinical Psychology (4th Edition).

41 Emotions, prefrontal cortex, and decision-making
- Eagleman (2015) The Brain The Story of You.
- Damasio (1994) Descartes' Error.
- Damasio (1999) The Feeling of What Happens.
- Bechara, H Damasio and A Damasio (2000) Emotion, Decision Making and the Orbitofrontal Cortex. Oxford Journals. Volume 10, Issue 3. Accessed on December 16, 2015. http://cercor.oxfordjournals.org/content/10/3/295.full
- Damasio (2003) Looking for Spinoza. Pages 141–145
- Lenzen (2005) Damasio interview—Feeling Our Emotions. Scientific American (March 24, 2005) http://www.scientificamerican.com/article.cfm?id=feeling-our-emotions Accessed on March 30, 2013.
- Antonio Damasio in Conversation with David Brooks, July 4, 2009, Aspen Ideas Festival, Aspen, CO. Uploaded on Aug 11, 2009. accessed on April 25, 2017. http://www.youtube.com/watch?v=1wup_K2WN0I.
- Damasio (2010) Self Comes to Mind.
- Bechara (2004) The role of emotion in decision-making. Brain and Cognition 55 (2004).
- Glimcher (2017) Understanding Human Decision-Making: Neuroeconomics. The Dana Foundation. Accessed on October 11, 2017. http://dana.org/ReportOnProgress/Glimcher/

- Booker (2014) "Study Cracks Brain's Emotional Code" Cornell Chronicle. July 9, 2014. As reported in Brain in the News, The Dana Foundation. Vol. 21, No. 8. September 2014. Page 1.
- Keltner, Oatley, Jenkins (2013) Understanding Emotions (Third Edition).
- Rolls, Grabenhorst (2008) The orbitofrontal cortex and beyond: From affect to decision-making. Progress in Neurobiology 86 (2008) 216–244. Accessed on December 9, 2015 http://www.utdallas.edu/~tres/plasticity2009/Rolls.pdf
- Miller, Freedman and Wallis (2002) The prefrontal cortex: categories, concepts and cognition. The Royal Society. Phil. Trans. R. Soc. Lond.
- Loewenstein and Lerner (2003) The Role of Affect in Decision Making (Chapter 31), *Handbook of Affective Sciences* Edited by Davidson, Scherer, and Goldsmith.
- Feldman Barrett (2017) How Emotions Are Made.

42 Fredrickson (2009) Positivity: Top-Notch Research Reveals the 3 to 1 Ratio That Will Change Your Life.

43 Implicit and Explicit Explained
- Merriam-Webster's 11th Collegiate Dictionary—words implicit and explicit defined.
- Galbraith (1983) The Anatomy of Power. (implicit and explicit conditioning as a tool of power).

44 Amygdala
- Silvia (2017). How One Memory Attaches to Another. Scientific American. July 2017.
- Davidson, Begley (2012) The Emotional Life of Your Brain.
- National Institute of Mental Health (NIMH) "Brain Basics" accessed January 17, 2014 http://www.nimh.nih.gov/health/educational-resources/brain-basics/brain-basics.shtml
- Carter (2009) The Human Brain Book.
- Edwards (2005) The Amygdala. The Dana Foundation. accessed May 14, 2015 http://www.dana.org/BrainWork/2005/The_Amygdala_The_Body's_Alarm_Circuit/
- Sukel (2013) Brain Reacts Differently to Internal vs. External Threats. The Dana Foundation. Accessed May 14, 2015. http://www.dana.org/News/Details.aspx?id=43269
- Schnabel (2009) Prefrontal Connection May be Key in Controlling Anxiety. The Dana Foundation. Accessed on May 14, 2015. http://www.dana.org/News/Details.aspx?id=43024.
- LeDoux (2006) The Power of Emotions. The Dana Foundation. accessed on May 14, 2015 http://www.dana.org/Publications/ReportDetails.aspx?id=44196.
- McGaugh (2003) Memory and Emotion.
- Linden (2011) The Compass of Pleasure.
- Hanson (2013) Hardwire Happiness.
- Keltner, Oatley, Jenkins (2013) Understanding Emotions (Third Edition).
- Merriam-Webster's 11th Collegiate Dictionary.
- American Heritage College Dictionary (2002) Houghton Mifflin Company.

- Trafton (2012) Patterns of connections reveal brain functions. MIT News Office. Accessed April 8, 2015 http://newsoffice.mit.edu/2012/face-recognition-0103.

45 Hippocampus
- Silvia (2017). How One Memory Attaches to Another. Scientific American. July 2017.
- Squire and Kandel (2000) Memory: From Mind to Molecules.
- Greene (2010) Making Connections. Scientific American Mind. July 2010.
- Kheirbek and Hen (2014). Add neurons, subtract anxiety. Scientific American. Accessed April 9, 2015 http://www.ncbi.nlm.nih.gov/pmc/articles/PMC4165637/
- Crist (2015) Old Neurons, New Tricks. Scientific American Innovators: Memory. Topix Media Lab Special #2, 2015.
- Byrne (2015) Chapter 7: Learning and Memory. Open-Access Neuroscience Electronic Textbook. Department of Neurobiology and Anatomy, The UT Medical School at Houston. http://neuroscience.uth.tmc.edu/s4/chapter07.html Accessed April 17, 2015.
- Richmond and Nelson (2007) Accounting for change in declarative memory. National Center for Biotechnology Information. accessed January 28, 2014. http://www.ncbi.nlm.nih.gov/pmc/articles/PMC2094108/
- Knierim (2007) The Matrix in Your Head. Scientific American Mind. June/July 2007.
- The Nobel Foundation (2014) Nobel Prize in Physiology or Medicine 2014 Press Release announcement (2014–10–06). Accessed on October 12, 2014. http://www.nobelprize.org/nobel_prizes/medicine/laureates/2014/press.html
- McGaugh (2003) Memory and Emotion.
- Keltner, Oatley, Jenkins (2013) Understanding Emotions.
- Patoine (2007) What's New in Neurogenesis: An Interview with Fred H. Gage, Ph.D. March, 2007. The Dana Foundation. Accessed April 19, 2015. http://www.dana.org/Publications/ReportDetails.aspx?id=44203
- Sorrells (2018) Human hippocampal neurogenesis drops sharply in children to undetectable levels in adults. Nature. Accessed March 8, 2018. https://www.nature.com/articles/nature25975.
- Merriam-Webster's 11th Collegiate Dictionary.
- American Heritage College Dictionary (2002) Houghton Mifflin Company.

46 2 percent body mass 20 percent blood flow
- Goldman (2016) Brain. Merck Manual—Consumer Version. Accessed October 10, 2016 http://www.merckmanuals.com/home/brain,-spinal-cord,-and-nerve-disorders/biology-of-the- nervous-system/brain
- RandomHistory.com (2010) 36 Interesting Facts About . . . The Human Heart.—Posted January 28, 2010. http://facts.randomhistory.com/human-heart-facts.html
- Lehnardt (2016) 41 Interesting Facts about the Human Heart. Fact Retriever LLC. https://www.factretriever.com/human-heart-facts. Published November 28, 2016. Accessed on January 15, 2018.
- Feldman Barrett (2017) How Emotions Are Made. page 57.

[47]   Pattern of Frequently Repeated Messages
-   Kahneman (2011) Thinking, Fast and Slow. page 62.
-   Perry, Szalavitz (2006) The Boy Who Was Raised as a Dog. Page 27, 46.

[48]   Sapolsky (2017) Behave. The Biology of Humans at Our Best and Worst. Chapter 11 Us Versus Them.

[49]   Affinity Bias
-   Merriam-Webster's 11th Collegiate Dictionary
-   Porges (2007). The Polyvagal Perspective. Accessed July 30, 2015. http://www.ncbi.nlm.nih.gov/pmc/articles/PMC1868418/
-   Porges (2011). The Polyvagal Theory.
-   McArdle (2020) How to explain systemic racism to non-liberals like me. The Washington Post. Accessed July 17, 2020. https://www.washingtonpost.com/opinions/in-the-covid-19-world- systemic-racism-is-deadly/2020/07/14/aabe1672-c601–11ea-b037-f9711f89ee46 story.html

[50]   Confirmation bias
-   David (2017) "The Upside of Bad Moods." Time Special Edition: The Science of Emotions.
-   Lilienfeld and Byron (2013) Your Brain on Trial. Scientific American Mind. Jan/Feb 2013.
-   Fox (2013) The Essence of Optimism. Scientific American Mind. Jan/Feb 2013.
-   Sapolsky (2017) Behave.
-   ScienceDaily—Science Reference. accessed November 10, 2017. https://www.sciencedaily.com/terms/confirmation_bias.htm
-   Zweig (2009) How to Ignore the Yes-Man in Your Head. Wall Street Journal. Accessed Nov 10, 2017. http://online.wsj.com/article/SB10001424052748703811604574533680037778184.html

[51]   Merriam-Webster's 11th Collegiate Dictionary.

[52]   Garner (Editor) (2004) Black's Law Dictionary.

[53]   Reward Pathway.
-   National Institute on Drug Abuse (2007) The Neurobiology of Drug Addiction; Section II. The Neurobiology of Drug Addiction. Accessed on July 12, 2015. http://www.drugabuse.gov/publications/teaching-packets/neurobiology-drug-addiction
-   Linden (2011) The Compass of Pleasure. Pages 16, 18, 144.
-   Keltner, Oatley, Jenkins (2013) Understanding Emotions (Third Edition). Pages 146–147.
-   Giedd (2009) The Teen Brain: Primed to Learn, Primed to Take Risks. The Dana Foundation.
    Thursday, February 26, 2009. Accessed on June 9, 2015.
    http://www.dana.org/Cerebrum/2009/The_Teen_Brain_Primed_to_Learn,_Primed_to_Take_Risks/
-   Kluger (2012) "Getting to No." TIME Magazine—U.S. Edition—March 5, 2012. pages 42–47.
-   U.S. Department of Health and Human Services (HHS), Office of the Surgeon General, Facing Addiction in America: The Surgeon General's

Report on Alcohol, Drugs, and Health. November 2016. CHAPTER 2. THE NEUROBIOLOGY OF SUBSTANCE USE, MISUSE, AND ADDICTION. Accessed on November 17, 2016. https://addiction.surgeongeneral.gov/chapter-2-neurobiology.pdf

- National Institute of Drug Abuse (2019) Neurobiology of Drug Addiction; Section II: The Reward Pathway and Addiction; 5: Addiction. (last updated November 2019). Accessed on January 14, 2020. https://www.drugabuse.gov/publications/teaching-packets/neurobiology-drug- addiction/section-ii-reward-pathway-addiction/5-addiction

- American Psychiatric Association (2013) Diagnostic and Statistical Manual of Mental Disorders— Fifth Edition (DSM-5) [Substance-Related and Addictive Disorders, page 481–585 | Gambling Disorder (312.31) pages 585–589 | Internet gaming disorder, page 795].

54 Like\Want

- Taber, Black, Porrino, Hurley (2012) Neuroanatomy of Dopamine: Reward and Addiction. Journey Neiropsychiatry Clin Neurosci 24:1, Winter 2012. accessed August 12, 2020 https://neuro.psychiatryonline.org/doi/10.1176/appi.neuropsych.24.1.1?url_ver=Z39.88– 2003&rfr_id=ori%3Arid%3Acrossref.org&rfr_dat=cr_pub++0pubmed&

- Keltner, Oatley, Jenkins (2013) Understanding Emotions (Third Edition). Page 146.

55 Denworth (2015) The Social Power of Touch. Scientific American Mind. July/August 2015. Page 35.

56 Saunders and Richard (2011) Shedding Light on the Role of Ventral Tegmental Area Dopamine in Reward. The Journal of Neuroscience, 14 December 2011. Accessed on July 12, 2015. http://www.jneurosci.org/content/31/50/18195

57 Dopamine

- Carter (2009) The Human Brain Book. Glossary page 243.

- 60 Minutes—Morley Safer (correspondent), David Browning (producer). "Hooked: Why bad habits are hard to break" (interview with Dr. Nora Volkow, head of the National Institute on Drug Abuse). Aired April 29, 2012. http://www.cbsnws.com/8301–18560_162–57423321/hooked-why-bad-habits-are-hard-to- break/?tag=contentMain;cbsCarousel

- Saunders and Richard (2011) Shedding Light on the Role of Ventral Tegmental Area Dopamine in Reward. The Journal of Neuroscience, 14 December 2011. Accessed on July 12, 2015. http://www.jneurosci.org/content/31/50/18195

58 Ventral striatum

- Gregorios-Pippas, Tobler, Schultz (2014) Short-Term Temporal Discounting of Reward Value in Human Ventral Striatum. Journal of Neurophysiology. Published 1 March 2009. http://jn.physiology.org/content/101/3/1507.long Accessed on July 19, 2015.

- Taber, Black, Porrino, Hurley (2012) Neuroanatomy of Dopamine: Reward and Addiction.
Journey Neiropsychiatry Clin Neurosci 24:1, Winter 2012.

- Fisher (2016) Addicted to Food, Games, Sex, the Internet. Scientific American

Mind. January / February 2016. Page 46.

59   Kardaras (2016) Generation Z: Online and at Risk? For today's teens, more
      followers online may mean fewer friends in real life—and a path to behavioral and
      psychological problems later on. Scientific American Mind. September/October
      2016. Page 67.

60   Other an addiction problem
      Ultra-processed foods.
      - Monteiro, Cannon, Levy, et al. (2016) NOVA. The star shines bright. [Food
        classification system] January-March 2016. Accessed on January 14, 2020.
      -   http://archive.wphna.org/wp-content/uploads/2016/01/WN-2016–7–1–3–
        28–38-Monteiro- Cannon-Levy-et-al-NOVA.pdf
      - Rosenbloom (2018) What is ultra-processed food and how can you eat less of
        it? Heart and Stroke Foundation of Canada. Accessed on January 14, 2020.
        https://www.heartandstroke.ca/articles/what-is-ultra-processed-food
      - Harvard T.H. Chan School of Public Health (2020) Processed Foods and
        Health. The Nutrition Source. Accessed on January 14, 2020. https://www.
        hsph.harvard.edu/nutritionsource/processed-foods/
      - Smartphone addiction
      - Walton (2017) Phone Addiction is Real—and so are its Mental Health
        Risks. Forbes (December 11, 2017). https://www.forbes.com/sites/
        alicegwalton/2017/12/11/phone-addiction-is-real-and-so-are-its-mental-
        health-risks/#38d5d70a13df Reprint in "Brain in the News" The Dana
        Foundation. January 2018 (pages 1– 2).
      - Glei (2016) Why we're addicted to email. Time Magazine. October 10, 2016.
        page 23.
      - Komaroff (2016) Could my teen be addicted to her smartphone? Ask
        Doctor K. Harvard Health Publications. Accessed Dec 8, 2016. http://www.
        askdoctork.com/teen-addicted-smartphone-201609099470

61   Stages of Behavioral Change
      - Center of Excellence for Infant and Early Childhood Mental Health
        Consultation (2022) Toolbox Resources, Stages of Readiness for Behavior
        Change. Last Updated: 04/25/2022. Accessed on May 30, 2022. https://www.
        samhsa.gov/iecmhc/toolbox/communications.
      - LaMorte (2019) The Transtheoretical Model (Stages of Change). Boston
        University School of Public Health. Date last modified: September 9, 2019.
        Accessed on May 30, 2022. https://sphweb.bumc.bu.edu/otlt/mph- modules/
        sb/behavioralchangetheories/behavioralchangetheories6.html
      - Merriam-Webster's 11th Collegiate Dictionary

62   Substance Abuse and Mental Health Services (2012) SAMHSA's Working
      Definition of Recovery (10 Guiding Principles of Recovery). Accessed on June 25,
      2022. https://store.samhsa.gov/sites/default/files/d7/priv/pep12-recdef.pdf

63   Ingersoll, (1906) The life of animals; the mammals. The Great Saber-Toothed
      Tiger – Smilodon. Page 87. Accessed on April 13, 2020 https://babel.hathitrust.
      org/cgi/pt?id=hvd.32044107328510&view=1up&seq=105

64   Types of Traumatic Events
      - Institute of Medicine. 2003. Preparing for the Psychological Consequences of

Terrorism. Chapter: 2. Pages 43–44. Accessed April 26, 2019. https://www.nap.edu/read/10717/chapter/4
- Schiraldi (2009) The Post-Traumatic Stress Disorder Sourcebook. Table 1.1, page 5.

[65] Appel (1999) "Fighting Fear." American Heritage Magazine (October 1999)

[66] Adverse Childhood Experiences (ACEs)
- Center for the Application of Prevention Technologies (2017) Adverse Childhood Experiences. Substance Abuse and Mental Health Services Administration. Accessed on June 15, 2017. https://www.samhsa.gov/capt/practicing-effective-prevention/prevention-behavioral- health/adverse-childhood-experiences
- Centers for Disease Control and Prevention (2016) Adverse Childhood Experiences (ACEs). accessed June 15, 2017. https://www.cdc.gov/violenceprevention/acestudy/

[67] National Transportation Safety Board (2010) Railroad Accident Report: Collision of Metrolink Train 111 With Union Pacific Train LOF65-12 Chatsworth, California September 12, 2008.

[68] Concept of Trauma
- van der Kolk, Weisaeth and van der Hart. Chapter 3—History of Trauma in Psychiatry. Pages 47–67. In van der Kolk, McFarlane, Weisaeth (Editors) (1996) Traumatic Stress: The Effects of Overwhelming Experience on Mind, Body, and Society
- Substance Abuse and Mental Health Services Administration (2014) Concept of Trauma and Guidance for a Trauma-Informed Approach. page 7–8. https://store.samhsa.gov/shin/content/SMA14–4884/SMA14–4884.pdf. Accessed Sept 16, 2017.
- Substance Abuse and Mental Health Services Administration (2017) Resource Guide to Trauma- Informed Human Services. https://www.acf.hhs.gov/trauma-toolkit. Accessed Sept 15, 2017.
- American Psychiatric Association (2013) Diagnostic and Statistical Manual of Mental Disorders— Fifth Edition (DSM-5).

[69] Vicarious Trauma
- British Medical Association (2022) Vicarious trauma: signs and strategies for coping. Last reviewed: 17 January 2022. Accessed on March 16, 2022. https://www.bma.org.uk/advice-and- support/your-wellbeing/vicarious-trauma/vicarious-trauma-signs-and-strategies-for-coping
- Pearlman and McKay (2008) Vicarious Trauma. Headington Institute. accessed on April 30, 2022. https://www.headington-institute.org/resource/vicarious-trauma-what-can-managers-do/

[70] Trauma-Informed Care
- Perry and Winfrey (2021) What Happened to You?
- Center for Health Care Strategies (2021) What is Trauma-Informed Care? Accessed on June 1, 2022. https://www.traumainformedcare.chcs.org/what-is-trauma-informed-care/
- The Administration for Children and Families, the Substance Abuse and Mental Health Services Administrations, the Administration for Community

Living, the Offices of the Assistant Secretary for Health and the Assistant Secretary for Planning and Evaluation at U.S. Department of Health & Human Services. Resource Guide to Trauma-Informed Human Services. Accessed on June 1, 2022. https://www.acf.hhs.gov/trauma-toolkit

- Substance Abuse and Mental Health Services Administration (2015) Trauma-Informed Care in Behavioral Health Services; KAP Keys for Clinicians Based on TIP 57. Publication ID: SMA15-4420. Accessed on June 1, 2022. https://store.samhsa.gov/product/Trauma-Informed-Care-in- Behavioral-Health-Services/SMA15-4420

- Substance Abuse and Mental Health Services (2012) SAMHSA's Working Definition of Recovery (10 Guiding Principles of Recovery). Accessed on June 25, 2022. https://store.samhsa.gov/sites/default/files/d7/priv/pep12-recdef.pdf

71   Association feature of memory
- Kahneman (2011) Thinking, Fast and Slow
- Genova (2021) Remember: The Science of Memory and the Art of Forgetting.

72   Kahneman (2011) Thinking, Fast and Slow.

73   Mere Exposure Effect
- Kahneman (2011) Thinking, Fast and Slow.
- Fournier (2016) Mere Exposure Effect. Published on PsychCentral.com. Last reviewed: July 17, 2016. Accessed on January 5, 2017. http://psychcentral. com/encyclopedia/2009/mere- exposure-effect/
- Definition of mere exposure effect. Accessed on January 5, 2017. http://www.businessdictionary.com/definition/mere-exposure-effect.html

74   Kahneman (2011) Thinking, Fast and Slow.

75   United States Holocaust Memorial Museum. "Nazi Propaganda." Holocaust Encyclopedia. Accessed on June 11, 2022. https://encyclopedia.ushmm.org/content/en/article/nazi-propaganda. PRESS KIT—State of Deception. Accessed on January 15, 2017. listed as a retired exhibition as of June 11, 2022. https://www.ushmm.org/information/press/press-kits/traveling-exhibitions/state-of-deception

76   Power
- Merriam-Webster's 11th Collegiate Dictionary.
- Galbraith (1983) The Anatomy of Power.
- Center for Strategic and International Studies (2007) CSIS Commission on Smart Power.
- Nye (2004) Soft Power the Means to Success in World Politics.
- Nye (2008) The Powers to Lead.

77   Seitz, Ortutay (2020) Tech companies step up fight against bad coronavirus info. The Associated Press (April 15, 2020). Accessed on April 17, 2020. https://abcnews.go.com/Business/wireStory/tech-companies-step-fight-bad-coronavirus-info-70169688

78   Constitution for the United States of America, Amendment I (1791) America's Founding Documents; National Archives. Accessed on April 23, 2022. https://www.archives.gov/founding-docs

79   Garner (Editor) (2004) Black's Law Dictionary. Page 448.

80   Garner (Editor) (2004) Black's Law Dictionary. Pages 1239, 1220, 596, 1293.

81   Daniel Patrick Moynihan (March 16, 1927—March 26, 2003).

- Banaji and Greenwald (2013) BLIND SPOT. page xv Preface.
- Will (2010) The wisdom of Pat Moynihan. The Washington Post. October 3, 2010. Accessed on February 12, 2017. http://www.washingtonpost.com/wp-dyn/content/article/2010/10/01/AR2010100105262.html
- Daniel Geary (2015) The Moynihan Report: An Annotated Edition. The Atlantic (September 14, 2015). Accessed on February 12, 2017. https://www.theatlantic.com/politics/archive/2015/09/the-moynihan-report-an-annotated-edition/404632/

82   Parker (2016) "You Can Fool All the People": Did Lincoln Say It? History News Network. February 14, 2016. Professor of History at Kennesaw State University. Accessed on September 19, 2018. https://historynewsnetwork.org/article/161924

83   Benjamin Franklin Quote
Social Studies for Kids. Sayings of Benjamin Franklin. Accessed on October 8, 2018.
http://www.socialstudiesforkids.com/articles/ushistory/benjaminfranklinquotes.htm

84   CONFLICT RESOLUTION

- Kneel (1999) The Keys to Conflict Resolution.
- Menkel-Meadow, Love, Schneider (2006) Mediation: Practice, Policy, and Ethics.
- Herman (1992) Trauma and Recovery.
- Rodriguez (2015) Hit the Gym After Studying. Scientific American Mind. May/June 2015.
- McGaugh (2003) Memory and Emotion.

85   Perry, Szalavitz (2006) The Boy Who Was Raised as a Dog. Page 93.

86   Tit for Tat Strategy

- Platt (1973) Social Traps (Prisoner's Dilemma) American Psychologist, August 1973. page 645.
- Dawkins (1976, 1989, 2006) The Selfish Gene. Oxford University Press. see Chapter 12 (Prisoner's Dilemma).

87   LEVELS OF ORGANIZED HUMAN ACTIVITY

- Nano comes from "Nanometer" one billionth of a meter (Merriam-Webster's 11th Collegiate Dictionary).
- Micro and Macro
    o   Barker (1987) The Social Work Dictionary.
    o   Meenaghan (1987) Macro Practice: Current Trends and Issues. Encyclopedia of Social Work.
- Meta
    o   Diamond (1999) Guns, Germs, and Steel. TABLE 14.1.
    o   Duffy (2017) 9 Questions interviewing Klaus Schwab. Time Magazine. Jan 23, 2017.

88   Groupthink

- Merriam-Webster's 11th Collegiate Dictionary.
- Principal Investigators Association (2017) No. 109: Beware: Research Lab

Teams Are Not Immune to "Groupthink." https://principalinvestigators.org/no-109-beware-research-lab-teams- are-not-immune-to-groupthink/. Accessed on June 28, 2017.
   - Arnold (2011) Following the Crowd. Scientific American Mind. September/October, 2011.
   - Jason R. Wells (8/19/05) "Groupthink and the Challenger disaster." Accessed on June 28, 2017. http://www.wellsj.com/library/groupthink.shtml
   - MindTools.com (2017) "Avoiding Groupthink: Avoiding fatal flaws in group decision making" http://www.mindtools.com/pages/article/newLDR_82.htm. Accessed June 28, 2017.

[89] Human History of Settlement
   - Smith (1995) The Emergence of Agriculture.
   - Diamond (1999) Guns, Germs, and Steel.
   - American Museum of Natural History (2022) OLogy: The Science Website for Kids. Accessed on March 26, 2022. https://www.amnh.org/explore/ology

[90] Jet Propulsion Laboratory—California Institute of Technology
   - (1996) Solar System Portrait—Earth as 'Pale Blue Dot.' Addition Date: 1996–09–12. Accessed on April 30, 2022. Catalog Page for PIA00452 (nasa.gov). https://photojournal.jpl.nasa.gov/catalog/PIA00452
   - Interstellar Mission of Voyager 1 and Voyager 2. Accessed on April 30, 2022. https://voyager.jpl.nasa.gov/mission/interstellar- mission/#:~:text=Voyager%20 2%2C%20which%20is%20traveling,on%20until%20at%20 least%202025.

[91] Examples of Other Systems of Prejudice
Nazi ideology in Germany (1933–1945) started with The Enabling Act
   - United States Holocaust Memorial Museum (2020)"The Enabling Act." Holocaust Encyclopedia. https://encyclopedia.ushmm.org/content/en/article/introduction-to-the-holocaust. Accessed on June 18, 2020.
   - Bitesize Daily Lessons (2020) Hitler takes political control 1933–1934. BBC Part of History, Life in Nazi Germany, 1933–45. Accessed on June 18, 2020. https://www.bbc.co.uk/bitesize/guides/zxs2pbk/revision/3
   - Apartheid in South Africa (1948–1994)
   - BBC News Services (4 April 2018) South Africa profile—Timeline, Apartheid set in law. Access Date: October 27, 2021. https://www.bbc.com/news/world-africa-14094918
   - Knauer (2013) Nelson Mandela: A Hero's Journey. Time Special Commemorative Edition.
   - History.com Editors (2020) Apartheid. A&E Television Networks. Original Published Date: October 7, 2010. Last Updated: March 3, 2020. Access Date: June 18, 2020. https://www.history.com/topics/africa/apartheid
     BBC News Channel (2004) The Good Friday Agreement signed on 10 April 1998. Accessed on October 27, 2021. http://news.bbc.co.uk/2/hi/uk_news/northern_ireland/4079267.stm

[92] Fernyhough (2017) Talking to Ourselves. Scientific American. August 2017.

[93] Pattern of Frequently Repeated Messages
   - Kahneman (2011) Thinking, Fast and Slow. page 62
   - Perry, Szalavitz (2006) The Boy Who Was Raised as a Dog. Page 27, 46.

94 The Iceman

- South Tyrol Museum of Archaeology, Italy. Accessed on July 16, 2016. http://www.iceman.it/en/oetzi-the-iceman
- Sapolsky (2017) Behave. page 307—"Iceman" 1991 melting glacier . . . 5,300 years old.

95 Galbraith (1983) The Anatomy of Power.

96 Issues, Positions, and Interests

- Menkel-Meadow, Love, Schneider (2006) Mediation: Practice, Policy, and Ethics.
- Fisher and Ury with Patton (1981, 1991) Getting to Yes (second edition).
- Ury (2007) The Power of a Positive No.

97 Hans Selye (1907–1982)

- Canadian Medical Hall of Fame (2020) Hans Selye, MD PhD. Accessed on May 14, 2020. https://www.cdnmedhall.org/inductees/hansselye
- Rosch (2017) Reminiscences of Hans Selye, and the Birth of "Stress." The American Institute of Stress. https://www.stress.org/about/hans-selye-birth-of-stress. Accessed May 11, 2020.
- Sapolsky (2017) Behave. Appendix 2—the Basics of Endocrinology.

98 The American Institute of Stress. Accessed on May 10, 2020. https://www.stress.org/daily-life

99 Stress

- Sapolsky (2017) Behave. Chapter Eleven, Us Versus Them.
- van Der Kolk (2014) The Body Keeps Score.
- The American Institute of Stress (2014) 2014 Stress Statistics. accessed on August 5, 2017. https://www.stress.org/stress-research/
- Kraft (2006) Burned Out: Your Job is Extremely Fulfilling. Scientific American Mind. June/July 2006.
- Levitt (2018) Defines Toxic Stress. DNA Learning Center. accessed June 21, 2018. https://www.dnalc.org/view/1226-Toxic-Stress.html
- Stress
  - o Alice Park (2011) The Two Faces of Anxiety. TIME Magazine. December 5, 2011.
  - o Merriam-Webster's 11th Collegiate Dictionary.
  - o The American Institute of Stress. Accessed August 5, 2017 http://americaninstituteofstress.org/daily-life/
- SELYE (1976) Forty years of stress research. CMA Journal. Accessed August 5, 2017. http://www.ncbi.nlm.nih.gov/pmc/articles/PMC1878603/pdf/canmedaj01483–0055.pdf.

100 Office of the Surgeon General (2016) Facing Addiction in America. Accessed March 16, 2020. https://addiction.surgeongeneral.gov/

101 Pareto Principle

- Mackenzie (1975) The Time Trap; How to Get More Done in Less Time.
- Kume (1987) Statistical Methods for Quality Improvement.

102 Forgiveness

- McCullough (2008) Beyond Revenge.
- Schiraldi (2009) The Post-Traumatic Stress Disorder Sourcebook.

- Ducharme (2019) Want to Stay Healthy as You Age? Let Go of Anger. Time Magazine. May 2019.

103 Jefferson, Count Ten

- Thomas Jefferson Encyclopedia, Canons of Conduct. accessed on June 13, 2022 https://www.monticello.org/site/research-and-collections/canons-conduct
- Jefferson (1825) Letter from Thomas Jefferson to Thomas Jefferson Smith, February 21, 1825, with Poem and Decalogue. Library of Congress, Manuscript Division: The Thomas Jefferson Papers. accessed on June 13, 2022. https://www.loc.gov/resource/mtj1.054_1267_1268/?sp=2

104 Procedural Memory (Muscle Memory)

- Amso (2017) "When Do Children Start Making Long-Term Memories?" Scientific American Mind, Ask the Brains. Volume 28, Number 1, January/ February 2017.
- Eagleman (2015) The Brain: The Story of You.
- McGaugh and LePort (2015) Remembrance of All Things Past. Scientific American Innovators: Memory. Topix Media Lab Special #2, 2015.
- Gibb (2007) The Rough Guide to the Brain.
- Tononi and Cirelli (2013) Perchance to Prune. Scientific American. August 2013.
- Carter (2009) The Human Brain Book.

105 Emotional Intelligence

- Goleman, (1995) Emotional Intelligence. page 43.
- Ayan (2015) How to Control Your Feelings—and Live Happily Ever After. Scientific American mind. January / February 2015. Page 50.

106 Davidson, Begley (2012) The Emotional Life of Your Brain.

107 Ury (2007) The Power of a Positive No. Page 30–31, 174.

108 Positive and Negative emotions.

- Fredrickson (2003) "The Value of Positive Emotions. The Scientific Research Society. American Scientist. 2003 July–August.
- Feldman Barrett (2007) The Science of Emotion. National Research Council. 2008. Human Behavior in Military Contexts. https://doi.org/10.17226/12023. Accessed on June 1, 2014. www.nap.edu http://www.nap.edu/openbook.php?record_id=12023&page=R1
- Fredrickson (2009) Positivity (3 to 1 Ratio). Pages 39–48.
- Davidson, Begley (2012) The Emotional Life of Your Brain.
- Fredrickson (2013) Positive Emotions Broaden and Build. Accessed on March 2, 2013 http://www.unc.edu/peplab/publications/Fredrickson_AESP_final.pdf
- Fredrickson (2013). Updated Thinking on Positivity Ratios. American Psychologist. Accessed on April 15, 2014. http://www.unc.edu/peplab/ publications/Fredrickson%202013%20Updated%20Thinking.pdf

109 Fredrickson (2009) Positivity: Top-Notch Research Reveals the 3 to 1 Ratio That Will Change Your Life. The Positive Emotions Concept is based on Barbara Fredrickson's broaden-and-build theory.

110 Damasio (2010) Self Comes to Mind. Pages 119–120

111 Substance Abuse and Mental Health Services Administration (2013) Tips for Survivors of a Disaster or Other Traumatic Event: Managing Stress. Publication

ID: SMA-13-4776. accessed on May 24, 2022. https://store.samhsa.gov/sites/default/files/d7/priv/sma13-4776.pdf

112   van der Kolk, Weisaeth and van der Hart. Chapter 3—History of Trauma in Psychiatry. Pages 47–67. van der Kolk, McFarlane, Weisaeth (Editors) (1996) Traumatic Stress: The Effects of Overwhelming Experience on Mind, Body, and Society.

113   Denworth (2017) The Good and Bad of Empathy. Scientific American (December 2017).

114   Merriam-Webster's 11th Collegiate Dictionary.

115   Emotional Intelligence
       - Goleman, (1995) Emotional Intelligence.
       - Graves (2017) Unlock Your Emotional Intelligence. Time Special Edition: *The Science of Emotions*. Lisa Lombardi, Editor. Pages 10–13.
       - Yale Center for Emotional Intelligence. http://ei.yale.edu/. Accessed on January 16, 2018.

116   Davidson, Begley (2012) The Emotional Life of Your Brain.

117   Three Layers of Empathy
       - de Waal (2012) "The Antiquity of Empathy." Science 18 May 2012. Accessed December 30, 2017. http://science.sciencemag.org/content/336/6083/874.full
       - Denworth (2017) The Good and Bad of Empathy. Scientific American (December 2017).
       - Breheny Wallace (2017) Being empathetic is good, but it can hurt your health. Washington Post (Health & Science). September 25, 2017. Accessed on January 16, 2018 via the Yale Center for Emotional Intelligence.

118   Nickerson (2021) Emotional Contagion. Simply Psychology. published Nov 08, 2021. Accessed on April 6, 2022. https://www.simplypsychology.org/what-is-emotional-contagion.html

119   Hatfield, Rapson, Le (2013) Emotional Contagion and Empathy. *The Social Neuroscience of Empathy* by Jean Decety and William Ickes; Part II Social, Cognitive, and Developmental Perspectives on Empathy; Chapter 2. Print publication date: 2009. Accessed on-line January 10, 2018. https://books.google.com/books?hl=en&lr=&id=KLvJKTN_nDoC&oi=fnd&pg=PA19&dq=emotional+contagion+and+empathy&ots=gE81cYjg-0Z&sig=QdneMq3eNOrQ53IyNNh81cBiBUs#v=onepage&q=emotional%20contagion%20and%20empathy&f=false.

120   de Waal (2012) "The Antiquity of Empathy." Science 18 May 2012. Accessed December 30, 2017. http://science.sciencemag.org/content/336/6083/874.full

121   Peer Support
       - Flannery (1995) Violence in the Workplace. Chapter 7 "Employee Victim Debriefing."
       - Flannery (1998) Assaulted Staff Action Program.
       - Solomon (2004) Peer Support/Peer Provided Services. Psychiatric Rehabilitation Journal. Spring 2004. http://164.156.7.185/parecovery/documents/Solomon_Peer_Support.pdf
       - Substance Abuse and Mental Health Services Administration (2015) "Peer

Support and Social Inclusion." https://www.samhsa.gov/recovery/peer-support-social-inclusion. Last Updated: 07/02/2015. Accessed on January 14, 2018.
- Substance Abuse and Mental Health Services Administration (SAMHSA) (2017) "Who Are Peer Workers?" https://www.samhsa.gov/brss-tacs/recovery-support-tools/peers. Last Updated: 11/20/2017. Accessed on January 14, 2018.
  Substance Abuse and Mental Health Services (2012) SAMHSA's Working Definition of Recovery (10 Guiding Principles of Recovery). Accessed on June 25, 2022. https://store.samhsa.gov/sites/default/files/d7/priv/pep12-recdef.pdf

[122] Suddendorf (2018) Two Key Features Created the Human Mind. Scientific American. Sept 2018.

[123] Edwards and Luscombe. (2018) United by Grief. Time Magazine. December 10, 2018.

[124] Based on Vernon Joseph Baker's creed in life. (January 13, 1997) http://clinton6.nara.gov/1997/01/1997–01–13-president-remarks-at-medals-of-honor-presentation.html

[125] Concept of Cooperation and reciprocal altruism.
- Frankl (1984) Man's Search for Meaning; Experiences in a Concentration Camp. (Page 60).
- Alcock (2001) The Triumph of Sociobiology. Oxford University Press.
- Keltner, Oatley, Jenkins (2013) Understanding Emotions (Third Edition) page 252.
- Nowak (2012) Why We Help: The Evolution of Cooperation. Scientific American July 2012.
- Angier (2002) Why We're So Nice: We're Wired to Cooperate. New York Times. July 23, 2002.
- Merriam-Webster's 11th Collegiate Dictionary.

[126] Substance Abuse and Mental Health Services Administration (2015) Trauma-Informed Care in Behavioral Health Services; KAP Keys for Clinicians Based on TIP 57. Publication ID: SMA15–4420. Accessed on September 28, 2020. https://store.samhsa.gov/product/Trauma-Informed-Care-in-Behavioral-Health-Services/SMA15–4420

[127] Koocher and Keith-Spiegel (2008) Ethics in Psychology and the Mental Health Professions, Standards and Cases, Third Edition.

[128] Copernicus (1473—1543)
- Merriam-Webster's 11th Collegiate Dictionary.
- Danielson & Graney (2014) The Case against Copernicus: Copernicus famously said that Earth revolves around the sun. Scientific American. Volume 310, Issue 1. pages 72–77. http://sams.scientificamerican.com/article/the-case-against-copernicus/

[129] Zimmer (2004) Soul Made Flesh: The Discovery of the Brain—and How it Changed the World.

[130] Feldman Barrett (2017) How Emotions Are Made.

[131] Maslow's Hierarchy of Needs
- McLeod (2020) Maslow's Hierarchy of Needs. updated March 20, 2020. https://www.simplypsychology.org/maslow.html Accessed on March 30, 2020.

- Gordon (1977) Leader Effectiveness Training: L.E.T. Pages 22–23.
132   The Attachment: Human Needs list was influenced by:
- McNamara (2003) THE FOG OF WAR: An Errol Morris Film; Ten additional
  lessons from R.S. McNamara as DVD's special features.
- Substance Abuse and Mental Health Services Administration (SAMHSA),
  U.S. Department of Health and Human Services. Evidence based practices of
  Supported Housing and Supported Employment.
- Substance Abuse and Mental Health Services Administration (2022) Recovery
  and Recovery Support. Last Updated: 04/04/2022. accessed on June 25, 2022.
  https://www.samhsa.gov/find- help/recovery
133   Braum (1900) The Wonderful Wizard of Oz.
134   Neil A. Armstrong, the first man to walk on the moon (July 20, 1969)
- Kruesi (2019) The Moon, Our Lunar Companion. National Geographic. Page
  35, 36.
- National Aeronautics and Space Administration (2012) Biography of Neil
  Armstrong and NASA https://www.nasa.gov/mission_pages/apollo/apollo11.
  html. https://www.nasa.gov/centers/glenn/about/bios/neilabio.html. accessed
  on October 21, 2018. Bradley, Armstrong (2005) "Being The First Man On
  The Moon. First aired on November 6, 2005. 60 Minutes CBS News. Accessed
  on October 21, 2018. https://www.cbsnews.com/news/being- the-first-man-
  on-the-moon/
- *National Aeronautics and Space Administration*
- NASA JOHNSON SPACE CENTER ORAL HISTORY PROJECT ORAL
  HISTORY TRANSCRIPT with NEIL A. ARMSTRONG INTERVIEWED
  BY DR. STEPHEN E. AMBROSE AND DR. DOUGLAS BRINKLEY
- HOUSTON, TEXAS—September 19, 2001. Page 78 of 106 https://www.
  nasa.gov/sites/default/files/62281main_armstrong_oralhistory.pdf
- Lunar Exploration Timeline—Lunar Missions https://nssdc.gsfc.nasa.gov/
  planetary/lunar/lunartimeline.html
135   U.S. Navy Fact Sheet—Aircraft Carriers—CVN. Last Update: 9 June
  2016; Accessed on July 3, 2016 http://www.navy.mil/navydata/fact_print.
  asp?cid=4200&tid=200&ct=4&page=1 .
136   Damasio (2010) Self Comes to Mind. (pages 257, 269, 291).
137   U.S. Department of Health and Human Services (1999) Mental Health: A Report
  of the Surgeon General. Chapter 2: The Fundamentals of Mental Health and
  Mental Illness. Page 31
138   Affinity Bias
- Merriam-Webster's 11th Collegiate Dictionary.
- Sapolsky (2017) Behave. Chapter Eleven Us Versus Them. page 400.
- Porges (2007). The Polyvagal Perspective. Accessed July 30, 2015. http://www.
  ncbi.nlm.nih.gov/pmc/articles/PMC1868418/
- Porges (2011). The Polyvagal Theory.
- McArdle (2020) How to explain systemic racism to non-liberals like me. The
  Washington Post. Accessed July 17, 2020. https://www.washingtonpost.com/
  opinions/in-the-covid-19-world- systemic-racism-is-deadly/2020/07/14/
  aabe1672-c601–11ea-b037-f9711f89ee46_story.html

[139]   Availability bias
-   Kahneman (2011) Thinking, Fast and Slow. Page 138.
-   Banaji and Greenwald (2013) Blind Spot. Page 9.

[140]   Kahneman (2011) Thinking, Fast and Slow. Page 28.

[141]   BLUE LIE BIAS
-   Sapolsky (2017) Behave. page 393, 394, 395.
-   Scherer (2017) The State of Truthiness. Time Magazine. April 3, 2017. Page 39.
-   Smith (2017) How the Science of "Blue Lies" May Explain Trump's Support. Scientific American Weekly Review (email) Accessed March 30, 2017. https://blogs.scientificamerican.com/guest-blog/how-the-science-of-blue-lies-may-explain- trumps-support/?WT.mc_id=SA_WR_20170329.
-   YourDictionary definition. LoveToKnow Corp. http://www.yourdictionary.com/white-lie. Accessed April 11, 2017.
-   Merriam-Webster's 11th Collegiate Dictionary.

[142]   Confirmation bias
-   David (2017) "The Upside of Bad Moods." Time Special Edition: The Science of Emotions. Page 26.
-   Lilienfeld and Byron (2013) Your Brain on Trial. Scientific American Mind. Jan/Feb 2013. Page 47.
-   Fox (2013) The Essence of Optimism. Scientific American Mind. Jan/Feb 2013. Page 24.
-   Sapolsky (2017) Behave. page 403.
-   ScienceDaily—Science Reference. accessed November 10, 2017. https://www.sciencedaily.com/terms/confirmation_bias.htm
-   Zweig (2009) How to Ignore the Yes-Man in Your Head. Wall Street Journal. Accessed Nov 10, 2017. http://online.wsj.com/article/SB10001424052748703811604574533680037778184.html

[143]   Schacter (2001) The Seven Sins of Memory. page 139.

[144]   Ekman (2004) Emotions Revealed. page 181.

[145]   Schacter (2001) The Seven Sins of Memory. page 139, 151.

[146]   Kahneman (2011) Thinking, Fast and Slow. page 62

[147]   Schacter (2001) The Seven Sins of Memory. page 139.

[148]   Matilda Effect:
-   Steinmetz (2019) Where Credit is Due (Correcting the Record). Time Magazine. April 22, 2019.
-   Haynes (2017) What is the Matilda Effect. Center for Sustainable Nanotechnology. accessed August 10, 2020 http://sustainable-nano.com/2017/03/08/what-is-the-matilda-effect-and-how-can-we- improve-recognition-of-women-scientists/

[149]   Merriam-Webster's 11th Collegiate Dictionary

[150]   Saporito (2016) In a turbulent stock market, the best investment move is the least obvious. Time Magazine. March 28, 2016. Page 35.

[151]   Merriam-Webster's 11th Collegiate Dictionary

[152]   Stereotypical bias
-   Schacter (2001) The Seven Sins of Memory. page 139

- Kahneman (2011) Thinking, Fast and Slow.

153 Truth Bias
  - Schafer (2013) Truth Bias. Psychology Today. accessed March 2, 2019. https://www.psychologytoday.com/us/blog/let-their-words-do-the-talking/201306/truth-bias
  - Borenstein (2019) Science Says: People tend to believe informants like Cohen. Associated Press. accessed on March 2, 2019. https://www.apnews.com/22314724fd744dcfb9573a2214058f7c
  - Street, Masip (2015) The source of the truth bias: Heuristic processing? Scandinavian Journal of Psychology. June 2015, Pages 254–263. Accessed March 2, 2019. https://onlinelibrary.wiley.com/doi/pdf/10.1111/sjop.12204
  - Van Swol, Braun, Kolb (2013) Deception, Detection, Demeanor, and Truth Bias in Face-to-Face and Computer-Mediated Communication. Sage Journals. accessed March 2, 2019. https://journals.sagepub.com/doi/full/10.1177/0093650213485785

154 Merriam-Webster's 11th Collegiate Dictionary

155 Insula
  - Izuma (2015) "Ask the Brains" (Question: What happens to the brain when we experience cognitive dissonance?). Scientific American Mind. November/December 2015. page 72
  - Damasio (2010) Self Comes to Mind. Pages 83, 110, 124, 125, 126, 212
  - Davidson, Begley (2012) The Emotional Life of Your Brain. Pages 78—81
  - Iacoboni (2008) Mirroring People. page 119
  - Gu, Hof, Friston, and Fan (2013) "Anterior Insular Cortex and Emotional Awareness." The Journal of Comparative Neurology, Research in Systems Neuroscience. Accessed March 19, 2015. http://www.fil.ion.ucl.ac.uk/~karl/Anterior%20Insular%20Cortex%20and%20Emotional%20Awareness.pdf
  - Song, Jung, Jang, Kim, Shim, Park, Choi, and Kwon. (2011). Disproportionate Alterations in the Anterior and Posterior Insular Cortices in Obsessive–Compulsive Disorder. Public Library of Science. Accessed on April 2, 2015. http://www.ncbi.nlm.nih.gov/pmc/articles/PMC3141026/
  - Suzuki (2012) Emotional functions of the insula. Brain Nerve. accessed on March 19, 2015 http://www.ncbi.nlm.nih.gov/pubmed/23037601
  - Keltner, Oatley, Jenkins (2013) Understanding Emotions (Third Edition). page 134
  - Carter (2009) The Human Brain Book. Pages 22, 106, 126, 136, 137, 170, & 245.
  - Siegel (2007) The Mindful Brain. Pages 38, 103, 167, 169, 335
  - Hanson (2013) Hardwiring Happiness. Pages 11, 43, 44